Understanding Your Antenna Analyzer

A Radio Amateur's Guide to Measuring Antenna Systems

by Joel R. Hallas, W1ZR

Published by ARRL 100 YEARS

Cover Design: Sue Fagan, KB1OKW

Technical Illustration: David Pingree, N1NAS

Production: Jodi Morin, KA1JPA
Shelly Bloom, WB1ENT

Printed in USA

ISBN: 978-0-87259-288-9
First Edition
Second Printing

Table of Contents

Foreword

Designing, building and adjusting antennas are favorite pastimes of Amateur Radio operators. Arguably, an antenna analyzer of some type is the most important piece in the antenna oriented amateur's toolbox. With the appropriate analyzer at hand, the amateur can determine the details of an antenna system's characteristics and much, much more.

This book is intended to introduce readers with a basic understanding of radio equipment and antennas to the ins and outs of antenna analyzers. Subjects covered include an introduction to the various types of analyzers available, their component parts, how they operate and how to operate them to get the best data available. In addition to using analyzers for their primary function, the book discusses the ways some analyzers can be used as general purpose test instruments in the amateur's lab. The final chapter provides summary product reviews, with ARRL Lab data on 11 analyzers that typify the equipment currently available.

As with all ARRL books, you can check for updates and errata, if any, at **www.arrl.org/notes**.

David Sumner, K1ZZ
Executive Vice President
Newington, Connecticut
March 2013

Chapter 1

Why Measure Antennas?

Whether your antenna system is as grand as this one at the W1AW, the Maxim Memorial Station at ARRL headquarters, or is a stealthy piece of wire, your success on the air will depend in large measure on how you understand what it can do.

While some amateurs build their own transmitters and receivers, most leave that task to manufacturers. Most of us purchase radio equipment based on published specifications, use the equipment and, if we need or want something different, we trade in and start over.

Some amateurs view antennas in the same way. Even if this is the case, antenna performance and operation can be very dependent on surroundings as well as environmental degradation, so it is good to check operation from time to time. Most amateurs, on the other hand, build at least some of their antennas. Having measurement capability is almost essential in order to achieve the desired design goals.

Arguably antennas are the most important part of an amateur station in terms of operating success. The variation in primary performance characteristics between entry level and top of the line radio equipment is generally far less than the variation in performance among different antennas. In addition, the majority of amateur antennas can be built easily by the user at relatively low cost from readily available hardware — not something that can be said of most radio equipment (see **Figure 1.1**).

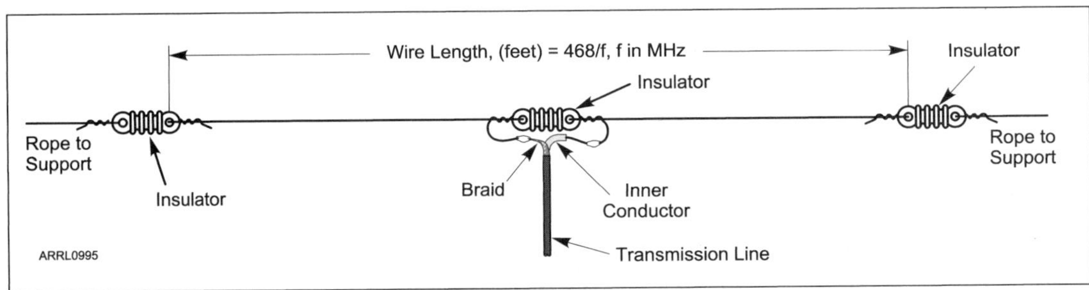

Figure 1.1 — A simple half wave dipole may be the most popular homebuilt antenna, but even this simple antenna can benefit from an antenna analyzer.

ANTENNA PARAMETERS TO BE MEASURED

There are many parameters related to antennas that we could measure. Not all can be measured by what we call SWR analyzers or antenna analyzers, but some can.[1] As we will discuss, not all antenna analyzers provide the same information.

[1]Many amateurs use the terms SWR analyzers or antenna analyzers interchangeably to describe the same type of device. Manufacturers may choose either label, and all such devices are really capable of more than either function, as we will discuss. We will use the term *antenna analyzer* in this book consistently to describe this equipment category, regardless of what parameters we are talking about analyzing.

Mechanical Parameters

Mechanical parameters of antennas are particularly important. These include the physical dimensions, critical to determining the space into which an antenna can fit, but there are others as well. The antenna weight is important in order to determine the design requirements of appropriate support structures. Equally important for some antennas is the exposed cross sectional area. This translates to horizontal wind load, often a limiting factor, particularly for rotary antenna support systems. No antenna analyzers can provide help here.

Radiation Performance Parameters

Key performance parameters of antennas include such characteristics as forward gain, front to back ratio (if applicable), pattern beamwidth and elevation pattern. Each of these is tightly coupled to the antenna surroundings and mounting arrangements, particularly the height above ground. Some antenna analyzers include a field strength measurement capability that could be helpful in evaluating some of these parameters, but that is not their main purpose.

Antenna Impedance Metrics

Antenna impedance measurement is where an antenna analyzer shines. The antenna analyzer is, at its roots, a device to measure antenna impedance or related quantities such as standing wave ratio (SWR). SWR is a measure of how well an antenna system is matched to provide a desired load, usually 50 Ω resistive. While basic analyzers may measure only SWR — often the most useful parameter — more advanced models may also indicate actual impedance, some the magnitude of the impedance, others the rectangular coordinates of the resistive and reactive components. We'll get into these concepts in the following chapters.

Chapter 2

Making Antenna Measurements

The Palstar PM2000A wattmeter is a typical station accessory that can be used to determine SWR based on forward and reflected power measurements while transmitting. It can be used in the antenna feed line during normal operation, but this type of meter has some functional limitations compared to an antenna analyzer.

There are many different types of devices that can provide information about an antenna system. Some devices are included as part of other equipment, some are laboratory-type instruments intended for indoor use, and some are portable units that can be taken outside and even carried up to the top of a tower.

Note that we refer to measurements about an *antenna system* rather than an *antenna* for a significant reason. In many cases, antennas are not in convenient locations to make measurements. If we could move an antenna close by, we would often change its characteristics due to a change in the proximity of the ground. The presence of the measurement equipment and person making the measurements can have an effect as well. Unless we are taking the measurement right at the antenna, we will measure the antenna characteristics as modified by the transmission line and any other media between the antenna and the measurement device.

STANDING WAVE RATIO MEASUREMENT

One of the most frequent measurements is of the standing wave ratio (SWR), fundamentally a transmission line measurement and not, strictly speaking, an antenna measurement. SWR is generally determined by measuring the ratio of reflected power to forward power. A transmission line terminated in its characteristic impedance (Z_0) will deliver all of the power to that matched load. Other than some reduction from line attenuation, with a matched load the magnitude of the voltage waveform will be the same anywhere along the line.

If the load is something other than equal to Z_0, there will be a signal reflected back from the load at a phase determined by the load impedance. The reflected wave will add and subtract from the incident wave over distance, resulting in the voltage being different as a function of the distance from the load. This voltage variation over distance is known as a standing wave.

The standing wave ratio is defined as the ratio of the maximum voltage on the line to the minimum voltage on the line. If the line is terminated in its Z_0, the maximum and minimum will be the same and the SWR will be 1:1. For some more background in SWR, see the sidebar "A Quick Discussion of Standing Wave Ratio" in this chapter.

In some equipment (many radio transceivers, for example) a meter selection to indicate SWR is provided. This reading is a measure of the match of the load to the transmitter's design impedance, often 50 Ω. If the transmitter is connected to its antenna system through a transmission line with a Z_0 of 50 Ω, the indicated reading will be the actual SWR on the transmission line as seen at the radio. If it is connected to anything else, the reading is really what the SWR would be if the load appeared at the transceiver end of a

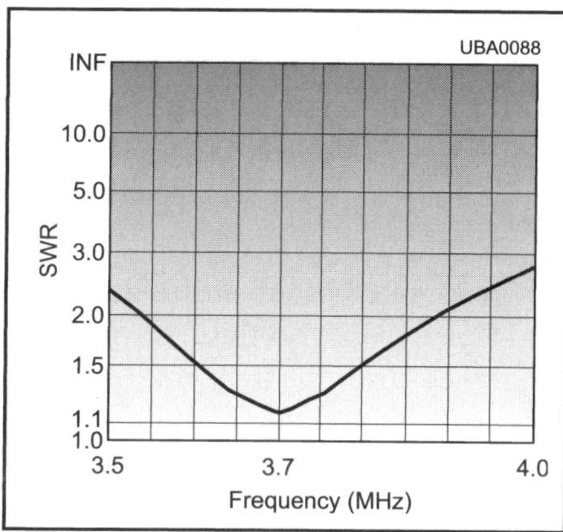

Figure 2.1 — The measured SWR of a representative wideband 80 meter antenna system tuned for 3.7 MHz.

50 Ω transmission line. This is a fine point, but one that's worth mentioning because some amateurs use 75 Ω coax to connect to their antennas.

For an antenna with a 50 Ω impedance at the desired frequency, a transmission line with a Z_0 of 50 Ω, and a transmitter that wants to see a 50 Ω load, in many cases it is sufficient to know that the SWR is 1:1, so all will work as designed. Of course most Amateur Radio operators want to operate over a band of frequencies, so a related set of data is the SWR over the desired portion of the band. That data can be compared to the transmitter SWR limit, often an SWR of 2:1 before the transmitter protection circuitry begins to reduce output power. **Figure 2.1** shows the measured SWR of a wideband 80 meter antenna system tuned for 3.7 MHz. As shown, the antenna will meet the 2:1 requirement over the range of 3.55 to 3.9 MHz, but not over the whole band.

Standing wave measurements used in transmission lines carrying RF power are often taken using line samplers that separately sample and measure the forward and reflected power. These are generally called *directional wattmeters* and are often found as standalone units, although some are built into transmitters. Some units provide just the two power readings, asking the user to use charts or formulae to determine the SWR (see **Figure 2.2**). Other units do the computation and read SWR

Figure 2.2 — The Coaxial Dynamics 81041 directional power meter has two line samplers, one for each direction. A switch selects between FORWARD and REFLECTED directions. Plug-in elements ("slugs") allow measurements across a wide range of frequencies and power levels. Coaxial Dynamics and Bird Electronics offer power meters that use a single slug, and direction is selected by turning the slug 180°.

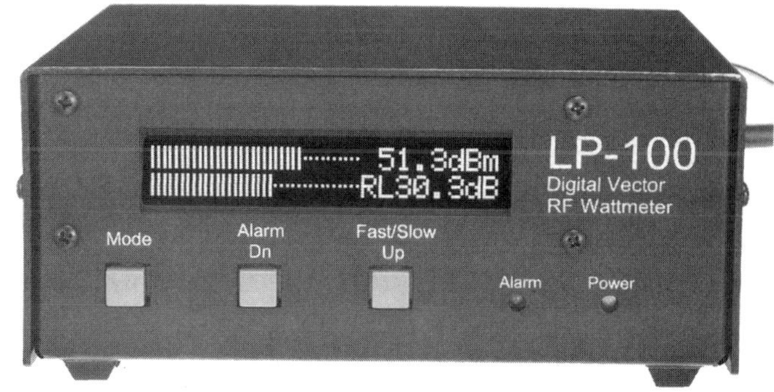

Figure 2.3 — The front panel readout of the Telepost LP-100 Digital Vector Wattmeter can display power in dBm and return loss in dB in addition to power in watts and SWR. In the vector mode, the display shows impedance. The top line is magnitude and phase, and the bottom line shows resistive and reactive components.

directly. More advanced units with additional processing provide detailed impedance data (see **Figure 2.3**). Lower power units sometimes use a Wheatstone bridge to balance the Z by calibrated variable resistive and reactive elements.

Impedance

The actual load impedance can be measured by some instruments. As noted previously, some advanced power meters can do this, as can other types of laboratory impedance bridges, if they cover the frequency range of interest. Some antenna analyzers can as well, with different units offering different forms of the data.

Field Strength

Unlike the previous measurements of SWR and impedance, the field strength is a measurement of the actual external effect of an antenna — measurement of the signal that is leaving the antenna for perhaps distant destinations. There are a wide range of devices for measuring field strength. The simplest type is basically an untuned crystal set receiver with a meter at its output (see **Figure 2.4**). A communications receiver with a signal strength meter (S meter) can serve as a kind of field strength meter, with the reference level at its antenna terminals, rather than in the area of the field. Professional quality field strength meters are essentially such receivers in combination with antennas of known aperture for different frequencies. Some antenna analyzers include a field strength measuring capability, generally close to the capabilities of the simplest type.

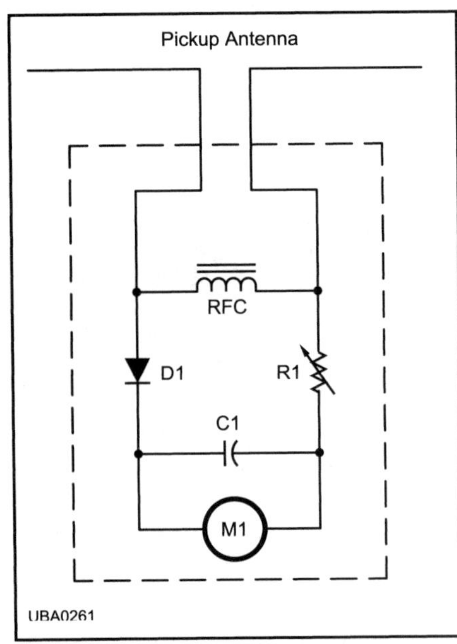

Figure 2.4 — Schematic diagram of a simple field strength meter. This type responds to all signals and has no direct calibration.

EQUIVALENT CIRCUIT OF AN ANTENNA

Of course, an antenna analyzer doesn't know what it is connected to. It could be an antenna system, but it could just as well be any load such as a resistor or a more complex network with capacitors or inductors. No matter what the load is, the information the analyzer gives us will be of the same form: the SWR as if it were a terminated transmission line, or the impedance as seen at its terminals.

To an analyzer, an antenna looks the same as some impedance value, generally a different impedance at

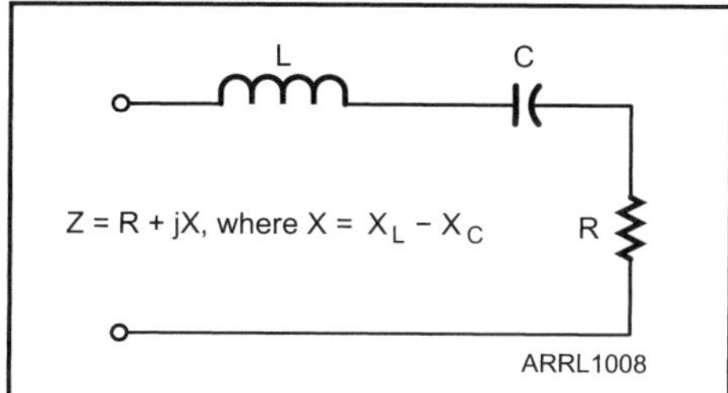

Figure 2.5 — To an antenna analyzer (or anything else connected to it), any antenna system impedance (Z) will look electrically equivalent to a resistor and either a capacitor or inductor. The values of resistance, inductance and capacitance will be different at different frequencies. Generally, at some frequencies the sign of the reactance will change, meaning that the impedance will change from inductive to capacitive or vice versa.

each frequency. Thus we can represent the load that the analyzer is connected to as a simple circuit consisting of a resistor and either a capacitor or inductor — an RLC network such as is shown in **Figure 2.5**.

If you're not used to the formula with the imaginary operator "*j*" in it, it just indicates that the impedance includes not only resistance but also the reactance in the form of inductive (+*j*) or capacitive (–*j*) reactance. At some frequency, the inductive reactance will equal the capacitive reactance and the result will be zero reactance (because they have opposite signs), leaving just resistance. This is called the network (or antenna system) resonant frequency. This is true for both antenna systems and our little RLC circuit. Note that at any frequency other than the resonant frequency, either the inductive or capacitive reactance will be greater and the net effect will be that of a two-element circuit with either a capacitor or inductor in series with the resistor.

WHAT ABOUT ANTENNA ANALYZERS?

Since this book is about antenna analyzers, it is probably time to introduce the concept. What makes an antenna analyzer different from the SWR measuring devices discussed so far is the way that it is packaged. It is a self-contained device that makes the SWR measurements (and possibly impedance measurements) that you might make with a power meter, but it makes them without the need for a radio transmitter. This is because an analyzer has a built in low-power signal source, along with the measurement and display subsystems, in a compact enclosure. As we'll discuss, different models offer quite different features, but they share this basic architecture. An analyzer offers some significant advantages over other measurement systems:

- While a transmitter-based measurement system can only be used on authorized operating frequencies, the antenna analyzer can be used to take data on any frequency within its operating range.
- The low signal intensity allows taking of data without causing interference to other users of the frequency — almost always.
- The self-contained, and often self-powered, device can be used easily

in the field, perhaps on the top of the tower, making antenna adjustments easier to accomplish.

As with all design choices, there are disadvantages as well.

• The most significant potential disadvantage is a downside of the low power measurement approach. If you have the analyzer connected to an antenna that happens to be picking up the signal from a strong nearby transmitter, such as an AM broadcast band transmitter, the coupled energy can interfere with the measurements. This is not a problem for most users, but it easily can be, particularly if you're in a metro area.

• Compared to most laboratory bench measurement systems, the information displays of compact handheld units are limited in size. This is usually not a major problem, but may be an issue in bright sunlight or other conditions. You just can't fit 10 pounds of display in a 5 pound bag!

What's in an Antenna Analyzer?

There is a wide range of features that may or may not be included in a given antenna analyzer design, explaining the perhaps 5:1 difference in prices among different models. **Figure 2.6** is a generic block diagram applicable to all antenna analyzers I can think of. The solid boxes and lines are absolutely required while the dashed lines represent components that may be included, depending on model (and price).

A very basic analyzer might consist of an analog tuning dial that is used to set the operating frequency and also provide a mechanical indication of

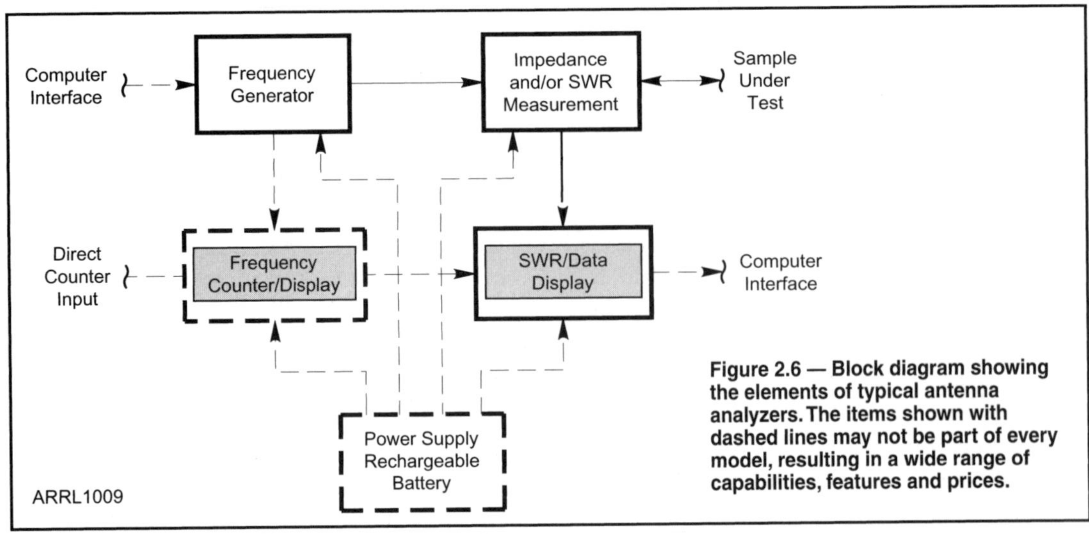

Figure 2.6 — Block diagram showing the elements of typical antenna analyzers. The items shown with dashed lines may not be part of every model, resulting in a wide range of capabilities, features and prices.

frequency. For a "data display" it will likely have an analog meter that will indicate just SWR. An ON-OFF switch disconnects the internal battery power supply. In spite of its simplicity, such a device can be surprisingly useful.

Below we will discuss each of the blocks and highlight some of the different features that may be associated with each block.

Frequency Generator. A major differentiator among units is the specified frequency range. Some units cover just MF through HF, perhaps 1 to 30 MHz, while others extend the range on each end — downward by perhaps a factor of 10 and upward by as much as a factor of 20 — perhaps resulting in a range from 0.1 to 600 MHz.

Obviously the complexity and cost of covering such a wide range will add considerably to the antenna analyzer's price, so it is good to have an idea of your requirements before you select a unit. In making your selection, keep in mind that the more flexible analyzers can be useful for tasks far beyond just measuring antennas. Virtually all can be used as bench signal generators, within limits that we will discuss.

Frequency Counter/Display. The accuracy and readout resolution of the generator may be equally important. The units with analog dials are most useful if a separate receiver is available to check their actual frequency. This may reduce a unit's utility for field operation. The more capable units indicate frequency by measuring it directly with a frequency counter, a worthwhile improvement for most applications. Be sure to look into the accuracy of the counter and its resolution (number of digits). Almost as important is the ability to set the frequency to any desired value and have it stay there. Being able to read to 1 Hz is not very useful if the tuning knob changes frequency so quickly that setting to within 50 kHz of the desired frequency is all you can do. Similarly, if it drifts that much between setting and reading the results, the resolution is of little benefit.

A major side benefit offered by some units with built in frequency counters is that they can be used to measure the frequency of other sources as well. If the unit can also serve as a standalone laboratory frequency counter, its value will be much higher, unless you already have one.

Impedance Measurement. As noted, the basic antenna analyzer measures standing wave ratio, an important parameter of an antenna system. In many cases, that's all that is needed or desired. Any more information is a distraction. On the other hand, if a new system is not well matched to the impedance its radio wants to see, the details of the actual impedance of the termination can aid in designing a matching network. If the measurement is taken at the antenna itself, knowing the impedance can indicate the direction in which adjustments to antenna dimensions need to be made. This distinction about measurement location will be discussed later.

Some analyzers display the information with an analog meter, and some use a digital display. In general, the analog display can often provide all the information needed, especially if the system is close to being matched. However, the digital display can provide much more useful data, especially in the case that the system is far from being matched.

In addition to SWR, some analyzers provide a measure of return loss (RL), an equivalent parameter that indicates the relative level of the reflected signal to the forward signal. Return loss is usually expressed in decibels. While not frequently encountered in the amateur world, return loss is commonly used in other realms.

Impedance data can consist of the magnitude of impedance, or the value of the resistive and reactive components — generally more useful. If both the magnitude and phase angle of the impedance are offered, that is equivalent to the information from rectangular resistive and reactive components. In some analyzers, in addition to the magnitude of the R and X values, the sign of the X is provided, indicating if the reactance is capacitive or inductive. That information is necessary to design a matching network, but sign can be determined indirectly, as we will discuss.

SWR/Data Display. The nature of the display system will make a considerable difference in the precision and resolution of the data, with digital systems generally having the edge. On the other hand, most people find it easier to make adjustments while watching the movement of the pointer of an analog meter. Some digital displays include a bar graph option for that purpose, and the bar graph can work well if it responds quickly enough.

A whole different dimension is provided by analyzers that show not just data at a particular frequency, but also graphical data as the frequency is swept across a range. This can be very useful for adjusting antennas or other networks, such as filters. Some units provide this capability on the small screen on the analyzer, while others allow connection to a PC with companion software for analysis and display on the computer screen. In some cases the data can be saved in the analyzer and examined in detail on the PC after coming down from the top of the tower.

Power Supply, Rechargeable Battery. For the casual user, there is not much wrong with running the analyzer from a bunch of alkaline AA batteries from the local drug store. Many units provide the capability to use rechargeable batteries and a separate power supply. This option looks more attractive after multiple trips down the tower and off to the drug store, although the alkaline batteries usually last longer than a charge on a rechargeable battery. Another consideration is the number of batteries the analyzer requires. There is a big difference between having to replace two batteries versus eight or so — it's pretty easy to have two in your pocket while up the tower!

It would seem that an ac supply would be beneficial, and it can be for some applications. Keeping the antenna analyzer isolated from the grounding arrangements of the power grid can be a benefit, especially if measuring loads on balanced transmission lines. Still, for many applications, having that wall transformer available, especially for extended lab testing, can be helpful.

A Quick Discussion of Standing Wave Ratio

Earlier in this chapter we noted that an antenna system with an impedance (Z) matched to the characteristic impedance (Z_0) of its transmission line has a standing wave ratio (SWR) of 1:1. That is the case in which there are no standing waves and all energy from the transmitter is delivered to the load with no reflections on the transmission line. This is a very straightforward situation that is easy to understand. Diving into the details of this subject can get lengthy, and there are entire books devoted to the topic.[1] Fortunately, we won't need to get that deeply involved to understand measuring antenna systems. Still, we need to have some idea about SWR. What does it mean? What are the consequences when and if it is a problem? How do we solve it?

Characteristic Impedance

If we were to connect a battery to a long transmission line with no load and monitor the current that would flow with a high speed oscilloscope, we would notice an interesting effect. The combination of the line's series inductance and shunt capacitance, and to a lesser effect the wire resistance, would result in an initial current that would flow even though there is an open circuit at the far end.

The current will continue to flow until the signal propagates to the end of the line and returns. When it reaches the end of the line (traveling at a somewhat reduced speed of light) a *reflected* wave of the opposite polarity will propagate back because there can be no current flow at the open circuit. When the reflected wave returns to the source end of the line, the combination of the forward and reflected wave will result in zero current — just what we would expect for an open transmission line.

The ratio of the applied voltage to initial current is an impedance and we call this the *characteristic impedance* or Z_0 of the transmission line. This is the current that would flow into an infinite length of line no matter what is at the far end. Note that we didn't say anything about signal frequency here. This is strictly a matter of the way the line is constructed, especially the capacitance and inductance distributed along its length. For the usual low loss line, ignoring resistance, we can determine Z_0 as follows:

$$Z_0 = \sqrt{L / C}$$

where L and C are the inductance and capacitance per unit length, typically per foot or per meter. These values can often be found in manufacturers' specification sheets.

The Matched Transmission Line

If, instead of having an infinite line, we put a resistor with a resistance equal to the line Z_0 on the far end of the line, it will absorb the power as it arrives and there will be no reflection. The voltage and current relationship at the source end of the line will appear as if it were the same resistor, just located at the source.

This is exactly what we wanted! If the source is a radio designed to drive a 50 Ω load, and Z_0 is 50 Ω, and our load is an antenna with an impedance of 50 Ω, we get just what we want. To our transmitter it appears as if the antenna is connected directly to the antenna terminal of the radio. There is no reflected power, the SWR is 1:1 and all is well with the world.

The Mismatched Transmission Line

There are an infinite number of cases in which a transmission line is terminated, not with its Z_0, nor with an open or short, but with some other impedance. This results in a reflection of a signal that is smaller than the signal that was sent down the line. It is important to note that this signal does not represent lost power because it is re-reflected from the transmitter back up to the line. It is true that the power delivered to the antenna will equal the forward power minus the reflected power, but the power is re-reflected at the transmitter so the numbers go up together. This is illustrated in **Table 2.A** for the case of perfect (lossless) transmission line, assuming the transmitter can put power into the SWR shown. These assumptions are discussed in the next section.

So What's the Big Deal?

The key is the two reasonable assumptions noted above. Let's discuss each:

• Transmission lines are not lossless, and their losses increase with increasing SWR as we will discuss. Whether or not that's a problem depends on the line type, length, frequency and SWR.

• Often a more serious problem is that somewhere in the range of SWR in Table 2.A, transmitters will reduce power ("fold back") to avoid damage due to the higher current or voltage at the final stage that results from the higher than specified SWR. Typical transmitters start to reduce power by the time SWR reaches 2:1.

Note that the second problem is not a fundamental issue of SWR itself, but rather a design choice made by transmitter designers. Still, for most transmitters it is a real consideration that we have to deal with.

What Kind of Load Results in High SWR?

Any load that is different from Z_0 of the transmission line will result in an SWR greater than 1:1. For the case of the frequently encountered 50 Ω coaxial cable, that means any load that is not 50 Ω *resistive*. This can mean a load that is resistive with a value of other than 50 Ω, or a load that is resistive, but that also has capacitive (–*j*X) or inductive (+*j*X) reactance, or any of an infinite number of combinations of the two. Some examples of loads with different SWRs are shown in **Table 2.B**.

Note that it is very easy to determine the SWR for the resistive case — it's just Z_0/R or R/Z_0 depending on whether the load is

Table 2.B
Impedance of Loads that Result in Different SWRs For Ideal 50 Ω Coaxial Cable

R (Ω)	*X (Ω)*	*SWR*
50	0	1:1
25	0	2:1
100	0	2:1
50	±35	2:1
30	±18	2:1
16.7	0	3:1
150	0	3:1
50	±58	3:1
30	±40	3:1
5	0	10:1
500	0	10:1
50	±142	10:1
250	±250	10:1

Table 2.A
Example of Net Transmitted Power with 100 W Transmitter versus SWR with Lossless Transmission Line

SWR	*Reflection %*	*Forward Power*	*Reflected Power*	*Antenna Power*
1:1	0	100	0	100
2:1	10	111	11	100
3:1	25	133	33	100

greater or less than Z_0.

Thus it is important to know not just the resistive part, but also the equivalent series inductive or capacitive reactance to determine the SWR.

Determining SWR from Impedance Data

While the case for resistive loads is simple, the case for loads with reactance or *complex* loads is, well, more complex. Still, there are at least three ways that I know of to determine the SWR of a complex load, not counting measuring with an SWR meter:

- Use the software *TLW* (*Transmission Line for Windows*) that comes packaged with recent editions of *The ARRL Antenna Book*.[2] If you plug in the R and X values and the appropriate transmission line Z_0, it provides the SWR at each end of the line. You can select which end of the line has the measured Z (input or load). This is very handy since it will also calculate the line loss.
- This may be the most simple calculation to make graphically with a Smith Chart.[3] Recall that if the Z is just resistive (X = 0), the SWR is either Z_0/Z or Z/Z_0, depending on whether the Z is lower or higher than the Z_0. Recall also that a circle on a Smith Chart represents constant SWR. Thus if you enter the Z on the Smith Chart and draw a circle centered on the chart center, it will show the transformed Z for any length of line. Either point at which the circle crosses the resistive axis can be used to calculate the SWR as described above.
- If you have neither computer nor Smith Chart, you can compute the SWR directly. This can be found, for example, in a book I had as a text many years ago, a classic by the late John Kraus, W8JK, *Antennas*. In the first edition it is in the appendix on page 507. The calculation is done in two steps:

1. Find the voltage reflection coefficient σ:

$$\sigma = \frac{Z - Z_0}{Z + Z_0}$$

Note that the Z is a complex number, so the calculation is a bit tedious.

2. Find the SWR

$$SWR = \frac{1 + |\sigma|}{1 - |\sigma|}$$

where $|\sigma|$ indicates the magnitude of the complex reflection coefficient, σ, found in step 1.

[1]W. Maxwell, W2DU, *Reflections*, available from CQ Communications at **www.cq-amateur-radio.com**.

[2]*The ARRL Antenna Book,* 22nd Edition, **www.arrl.org/shop.**

[3]Paper copies of 50 Ω Smith Charts are available from ARRL, **www.arrl.org/shop**.

Chapter 3

Information Available from an Antenna Analyzer

A look at the displays of these four antenna analyzers provides clues to the different kinds of information available from different analyzers.

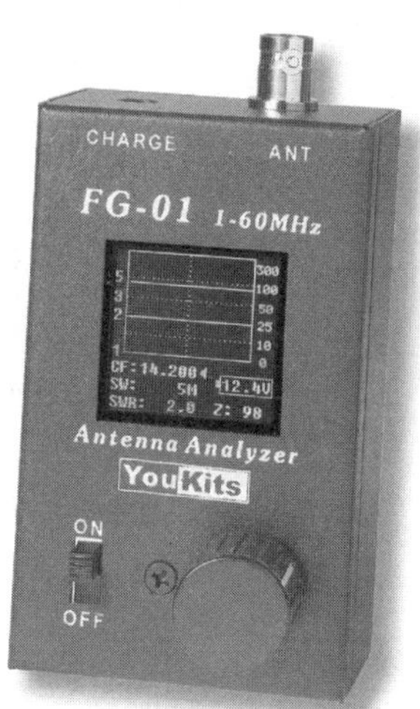

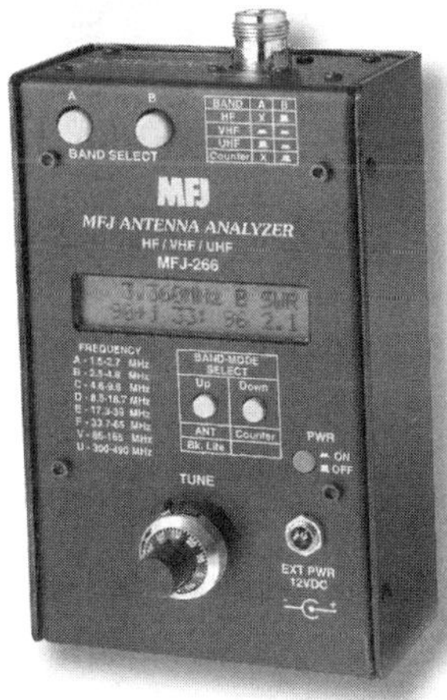

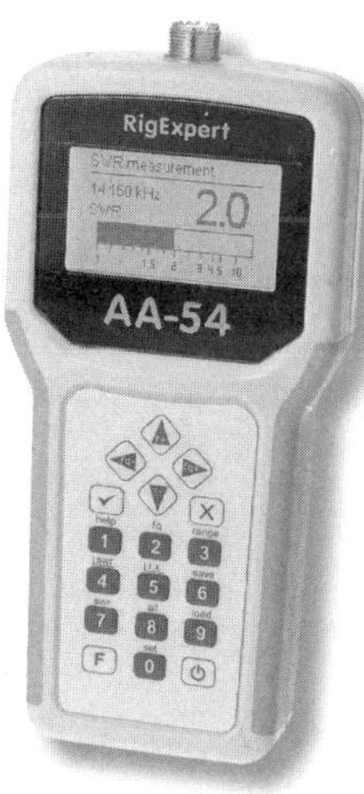

STANDING WAVE RATIO

Standing wave ratio (SWR) is the most common, and probably most useful, data available from an antenna analyzer. In fact, there are some analyzers that provide *only* SWR data. It is important in some circumstances to know that most, and perhaps all "antenna analyzers" do not measure standing wave ratio directly. After all, SWR is a transmission line effect and the analyzer is actually indicating data based on the measured impedance as seen at the measurement port as compared to the design impedance, usually 50 Ω.

This is not a particularly important distinction if the system you are measuring is designed to be a 50 Ω system. For example, with a resistive load of 50 Ω, we will measure a 1:1 SWR, as we would expect. That's true whether the load is a 50 Ω transmission line terminated in 50 Ω, or if it is a 50 Ω carbon (noninductive) resistor directly connected to the analyzer terminals. Similarly, if we have a resistive load of 100 Ω, we will measure a 2:1 SWR, and this will happen whether or not we have a 50 Ω transmission line between the analyzer and the resistor. On the other hand, if we have a perfectly matched 75 Ω transmission line, with a 1:1 SWR the antenna analyzer will indicate a 1.5:1 SWR. That indication corresponds to the ratio of 75 to 50, not the actual standing waves that may or may not exist on a transmission line that may or may not be there.

SWR is an important measure, especially if we are interested in measuring, or adjusting to, a 50 Ω load on the end of a 50 Ω transmission line. While this seems as if it would be a restrictive requirement, it is frequently the case we are dealing with. A typical application is the adjustment of the resonant frequency of perhaps a dipole antenna at a height at which it will have a resistive impedance of 50 Ω at some frequency. If we measure the SWR over a frequency range and find it has a 1:1 SWR at a different frequency, we can change its length to move its resonance to our desired frequency. If the impedance of the antenna is not 50 Ω resistive at any frequency, then it will be transformed by the mismatched transmission line to some other impedance at the analyzer. This can make analysis of the SWR complicated.

IMPEDANCE

As noted in a previous chapter, impedance is generally a complex value with both a resistive (real) and reactive (imaginary) component. There are two ways that antenna analyzers may display measured impedance values: in *rectangular* or *polar* coordinates. The alternatives are shown graphically in **Figure 3.1**. The measured impedance is equivalent to 40 Ω of resistance in series with 30 Ω of inductive reactance. The rectangular coordinates would

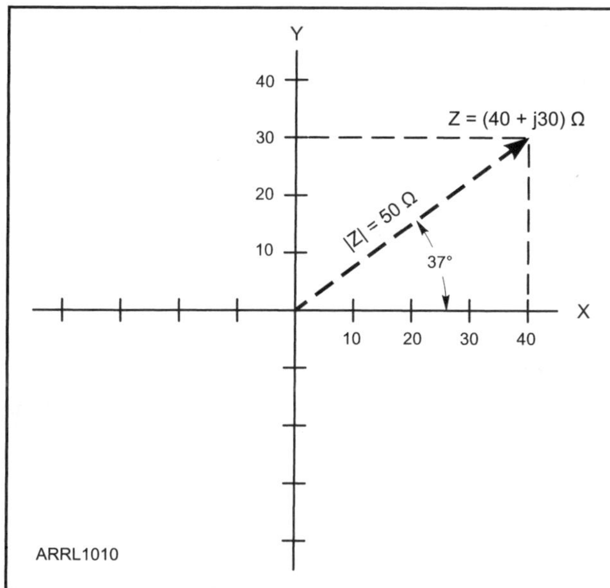

Figure 3.1 — Example of two ways to represent an impedance. Specifying the rectangular coordinates would express the resistive part and the reactive part, in this case 40 Ω and + *j*30 Ω respectively. An alternate presentation would be to show the magnitude of the impedance (50 Ω) and the phase angle (37°). They are equivalent, and either is correct. It's a matter of personal preference.

be expressed as Z = (40 + *j*30) Ω. To convert to polar coordinates, we find the magnitude (shown between parallel vertical bars) using the Pythagorean theorem, $|Z| = (X^2 + Y^2)^{0.5}$ and then the angle is just equal to $\tan^{-1}(Y/X)$. The Electrical Fundamentals chapter of *The ARRL Handbook* has more information on rectangular and polar coordinates (see the Radio Mathematics section).

RETURN LOSS

Some antenna system measurement equipment displays an alternate form of data that is equivalent to standing wave ratio, since they are both a function of the ratio of forward to reflected power. *Return loss* looks at the transmission line-antenna system as if it were a device that delivered back some of the power put into it.

The return loss of a transmission line with an infinite SWR is thus 1 or 0 dB, that is, all the power put into the "device" is returned as a reflection. The case of a matched system ends up looking the opposite way. That is because the system absorbs all the forward power and none is returned, so the return loss is infinite. Other than the fact that this may seem backward to those used to working with SWR, it is an equivalent measure.

One minor concern, depending on the nature of the measurement device, is that values of return loss for a system close to being matched are quite large. Because the values are expressed in decibels, this usually is not a major problem. For example, a 1.1:1 SWR is equivalent to return loss of 26.4 dB.

It is fairly easy to determine return loss from SWR in two steps, as follows:

1. Find the magnitude of the reflection coefficient (Γ).

$$\Gamma = \frac{SWR - 1}{SWR + 1}$$

Table 3.1
SWR, Reflection Coefficient and Return Loss

SWR (:1)	*Reflection Coefficient (Γ)*	*Return Loss (dB)*
1.05	0.0244	32.26
1.1	0.0476	26.44
1.2	0.0909	20.83
1.5	0.2000	13.98
1.7	0.2593	11.73
2.0	0.3333	9.54
2.5	0.4286	7.36
3.0	0.5000	6.02
3.5	0.5556	5.11
4.0	0.6000	4.44
4.5	0.6364	3.93
5.0	0.6667	3.52
7.0	0.7500	2.50
10.0	0.8182	1.74
15.0	0.8750	1.16
20.0	0.9048	0.87
25.0	0.9231	0.70
50.0	0.9608	0.35
100.0	0.9802	0.17

2. Determine the reflection loss in dB.

$$RL = -20 \times \log_{10}(\Gamma)$$

To determine the SWR from the reflection loss in dB:

$$SWR = \frac{1 + 10^{(RL/20)}}{1 - 10^{(RL/20)}}$$

Table 3.1 shows the values of return loss for a range of SWR readings. I found it easy to program the formulas into an *Excel* spreadsheet and then copy them to obtain the results shown. Tables and converters in either direction are also available on the Internet.

Chapter 4

Hooking it Up and Making it Play

Nancy, W1NCY, measures the SWR of her 2 meter and 70 cm dual band mobile antenna. Select an analyzer that measures the frequencies you need, or might need in the future.

Different analyzers have output data in different forms. Here we will discuss some of the data available while using an analyzer for antenna measurement. We'll look at what the data mean, how to interpret them and how to make use of them. Not all analyzers will provide all the types of data listed.

CONNECTING TO ANTENNA OR FEED LINE

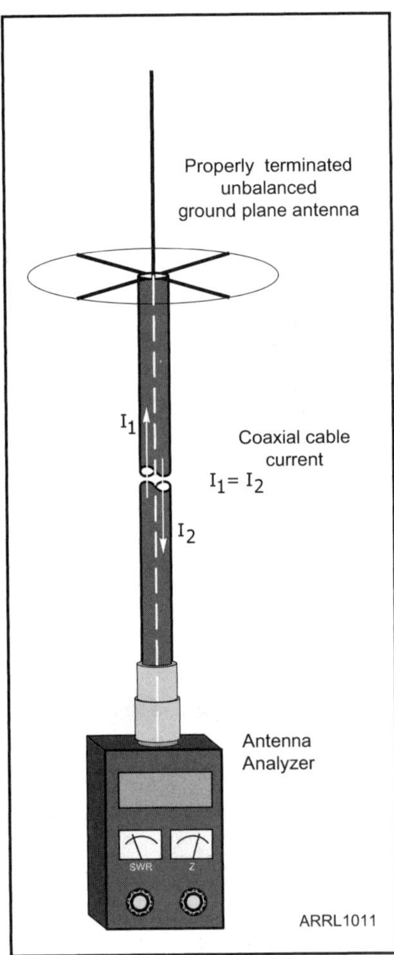

Figure 4.1 — Illustration of an antenna analyzer with an unbalanced coax output connected to an unbalanced antenna load through a length of unbalanced coax cable. Note that virtually all of the current is within the coaxial cable, as it should be for an accurate measurement.

The typical antenna analyzer has a coaxial connector on its top or side to connect to the load to be measured. Most analyzers designed for HF and VHF use a UHF (SO-239) socket, although BNC and Type N sockets are sometimes provided. If you are faced with connector incompatibility, a between-series adapter can be used. Keep in mind that the additional inch or so between the meter and the load acts like a short transmission line section and may be important to take into account in some applications, such as trimming feed lines for matching purposes.

Coax Shield Current Problems

Note that the typical analyzer is designed to interact with unbalanced coaxial cable in differential mode. If the load is not properly terminated to result in a differential mode with all currents inside the coax, the measurements may be inaccurate and dependent on the operator's body or power connection as much as the antenna system. **Figure 4.1** and **Figure 4.2** illustrate the concept.

In Figure 4.1, the analyzer is connected to a coaxial cable that is feeding a typical unbalanced load, in this case a ground plane antenna. The antenna currents will all return to the upper side of the ground plane, resulting in all of the current being within the coaxial cable. It is possible to induce current into the outside of the shield by improperly installing the antenna so that the antenna field is coupled to the coax. In most cases the cable can run directly downward for some distance to avoid such coupling.

In contrast, the arrangement of Figure 4.2 is similar, except in this case the coax is directly feeding an inherently balanced dipole. The coax shield is thick enough that skin effect keeps currents near the inner and outer surfaces. Thus the shield acts like two concentric conductors.

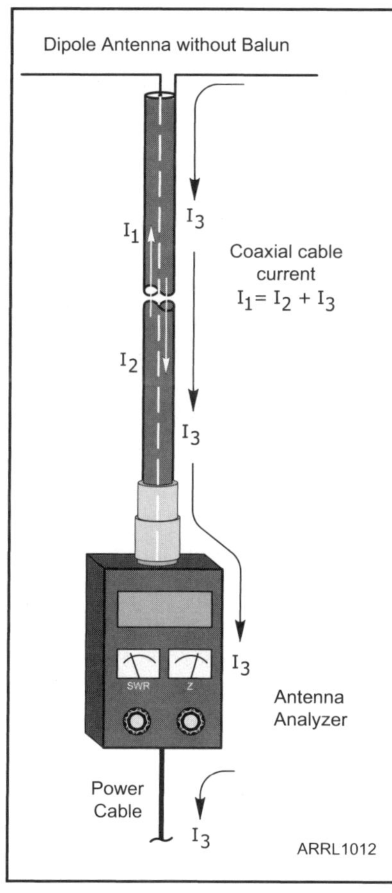

Figure 4.2 — Illustration of an antenna analyzer with an unbalanced coax output connected to a balanced antenna load through a length of unbalanced coax cable. Note that the shield current will split between half of the dipole and the outside of the shield resulting in readings of questionable value.

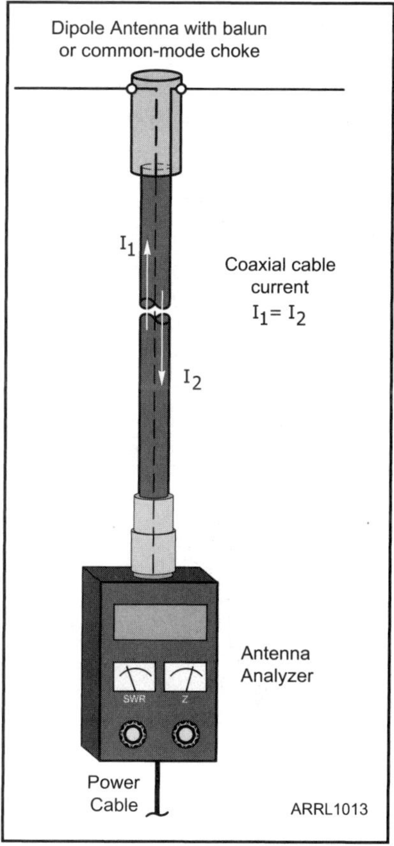

Figure 4.3 — The situation of Figure 4.2 modified through the insertion of a balanced-to-unbalanced transition (balun) or common mode choke to force the current to stay inside the coax. This avoids the problems of current on the outside of the coax. Any other effects of the balun must be taken into account in interpretation of the data.

If the shield is connected directly to part of the antenna as shown, the shield current will split between the dipole half and the outside of the coax shield. The division of current between the two conductors will depend on the relative impedances of the two paths, a function of the coax cable length and shield termination.

The resulting common mode current (as opposed to normal differential mode — between-the-conductors current) will make for inaccuracy in measurements of the antenna characteristics. It is also true that if connected to a radio, the same currents — much larger while transmitting — may result in problems with the radio equipment by making the equipment chassis *hot* with RF. This can result in curled moustaches while transmitting voice, as well as potential lockup of the transmit-receive switching process.

RF currents on the outside of the coax can also result in RF interference

(RFI) to telephones, alarm systems and other devices near the cable's path. During reception, undesired signals from household systems coupled to the shield will also be induced into the receiver, resulting in unnecessary noise and interference.

A way to eliminate the current on the outside of the coax is to insert a balanced-to-unbalanced transformer (balun) or common-mode choke at the termination, as shown in **Figure 4.3**. Avoiding coupling from the antenna to the transmission line is important in this case as well. This can generally be achieved by keeping the transmission line perpendicular to the antenna for as long a distance as possible. At least ¼ wavelength will usually provide satisfactory results.

An easy-to-make common mode choke can be formed by wrapping turns of the coaxial transmission line through a ferrite toroid. Using slip-on ferrite beads is another way to reduce current on the line. If you are trying to characterize the antenna itself, or trying for optimum operation, the choke should be at the antenna end of the line. On the other hand, if you are trying to characterize an antenna system that does not include a balun or choke, for the best characterization of the system connect the analyzer to the transmission line after the shield has been grounded at the bottom of the line, as it normally would be. Then place the balun or choke between the bottom of the line and the SWR analyzer. In that way the entire system, as normally used, will be analyzed.

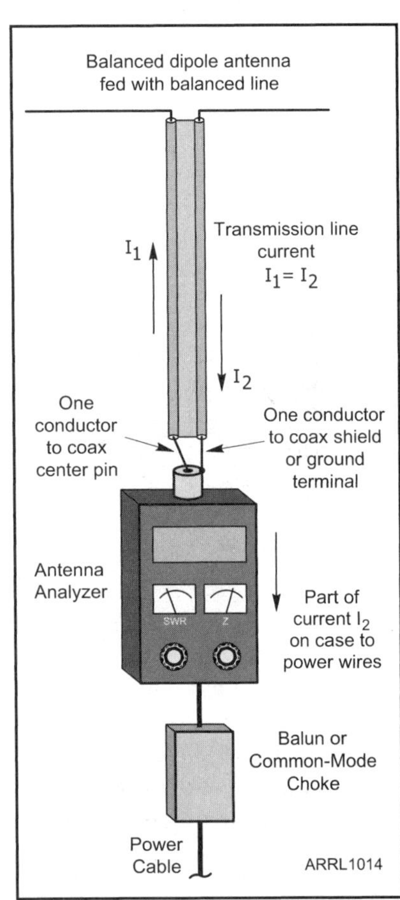

Figure 4.4 — The measurement of characteristics of balanced systems, such as balanced antennas fed by balanced lines, have effects similar to measuring balanced antennas fed with coax. A balun can be placed at the analyzer output, with the same concerns as above, or battery power can be used to avoid currents on the power wiring. Alternately, a choke can be located between the power wires and the analyzer as shown.

Measuring Balanced Lines

Balanced transmission lines, such as twinax, twisted pair, open wire line, window line and twinlead, can be connected to the analyzer in a number of different ways. Some analyzers have a ground terminal on the top of the case that can be used for one side of the line, with the other side inserted into the center pin of the coax connector. A banana plug makes a good fit to the center conductor of a UHF socket. See **Figure 4.4**.

Connecting the line as described above is feasible, however, it does tend to unbalance the line. To get as good a set of data as possible, use internal batteries rather than an external power supply, and take the readings with the analyzer positioned on an ungrounded surface. Any connections or contact to the analyzer will make the unbalance worse and will tend to invalidate the data.

A balun or common mode choke at the analyzer can eliminate the effects of this problem, but they become part of the measured system and the results must be analyzed and adjusted to move the effective measurement point to the balanced side of the balun. The use of coaxial cable wound on a toroid, as described above, is a good choice because the choke will not affect the measurement other than by adding the length of the coax wound in the choke. The signal under consideration is the differential one on the inside of the coax — not affected by the choke inductance.

An alternative is to wind turns of the dc power lead coming from the power supply through a similar toroid as close to the analyzer as possible. A *QST* author recommended using 15 turns on the toroid.[1] This isolates the power supply and ac wiring from the analyzer, although the problem with unbalance due to the analyzer itself is still present. This effect tends to be more significant as the frequency is raised, since the capacitive reactance on one side of the line is reduced with frequency.

MEASUREMENT FREQUENCY

While not generally an output parameter, the frequency that the data relates to is of critical importance to the relevance of all recorded data, so we list it first. Analyzers usually have a mechanical dial, a digital frequency readout or sometimes both.

An analyzer using a frequency control on an analog shaft with a mechanical pointer indicating the frequency does not directly provide enough resolution or accuracy by itself to yield meaningful data for many purposes. The solution is to couple the analyzer to a frequency counter to determine the actual frequency. This can be done to check on band edge locations before measurements are taken, or it can be used to check the frequency of minimum SWR after the data is taken. If a frequency counter is not available, a communications receiver with an accurate, high resolution frequency readout can be used to measure the frequency.

Digital frequency readouts generally provide more than sufficient precision to be able to perform most analyzer tasks, but don't let precision lull you into thinking that it equates to accuracy. If an internal frequency counter or direct digital synthesis (DDS) generator is employed, the accuracy will only be as good as the internal time base or frequency reference employed. In either case, it not likely to be of the same standard that we would expect from more serious test instruments. The remedy for the digital display is thus the same as that for the analog dial. Check the frequency against a known reference, using a frequency counter or communications receiver of accuracy appropriate to the accuracy needed for the task at hand. Unlike the analog dial units, once you have determined the accuracy or offset, it is unlikely to change a lot over time and an infrequent calibration check may be sufficient.

For either type of unit, it is important to determine the expected frequency drift and resetability. These are more likely to become important in the VHF range and above. Apply power to the analyzer and tune in its signal on a stable "warmed up" communications receiver. Check the frequency after 5, 10 and 30 minutes and record the drift. Repeat for other ranges.

MEASURING STANDING WAVE RATIO

SWR is generally displayed either on an analog meter or a digital display. The analog meter has a few advantages, such as being more visible in some kinds of bright lighting than some digital displays. In addition many amateurs find it easier to watch an analog meter while making adjustments, although some digital displays also include a fast moving bar graph that may be as good.

The digital display offers a number of advantages over an analog meter. Perhaps most obvious is increased precision of the measurement. Note that this is not the same as accuracy, and it is likely that the precision offered by a digital display exceeds the actual accuracy of SWR measurement. Still, precision can be important — especially if you are trying to notice small changes as frequency or antenna length is changed. While the analog meter may read to a few tenths, as in the difference between 2.2:1 and 2.4:1, the digital scale may indicate the difference between 2.20:1 and 2.21:1.

Perhaps more importantly in favor of the digital display is the displayed SWR range. The typical analog meter reads in a single scale that is compressed in the upper region. So while values around 2:1 or 3:1 are easily discerned, those at the higher range are not. For example, an analyzer with an analog display may be specified to measure from 1:1 to ∞, but in practice the range from 6:1 to ∞ may just be a line ½ inch long that makes reading an actual value impossible. I would call the *useful* range 1:1 to 6:1.

A unit with a digital display may show actual values up to 100:1 and then jump to ∞. Again, the accuracy probably doesn't match the precision, but the indication of direction of change with adjustments may be useful.

In addition to the analog meter and numerical digital display, some advanced units include a display screen. The RigExpert AA-54, for example offers a number of SWR display options. **Figure 4.5** is the AA-54 "basic" SWR display. Note that in addition to a digital SWR value, shown to hundredths, it indicates measurement frequency and provides a calibrated bar graph to facilitate making adjustments.

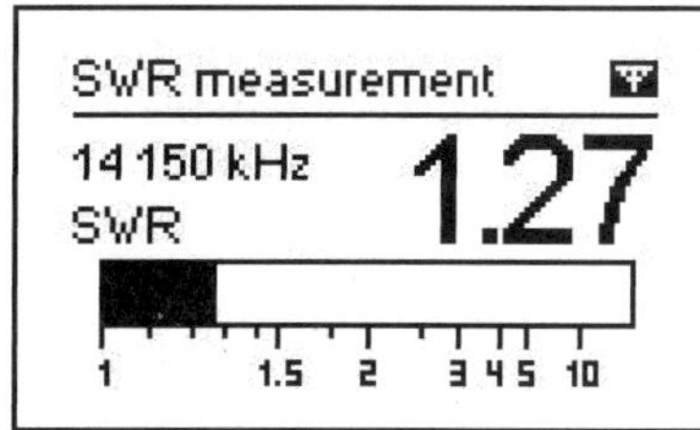

Figure 4.5 — The RigExpert AA-54 "basic" SWR display. Note that in addition to a digital SWR value, shown to hundredths, it indicates measurement frequency and provides a calibrated bar graph to facilitate making adjustments.

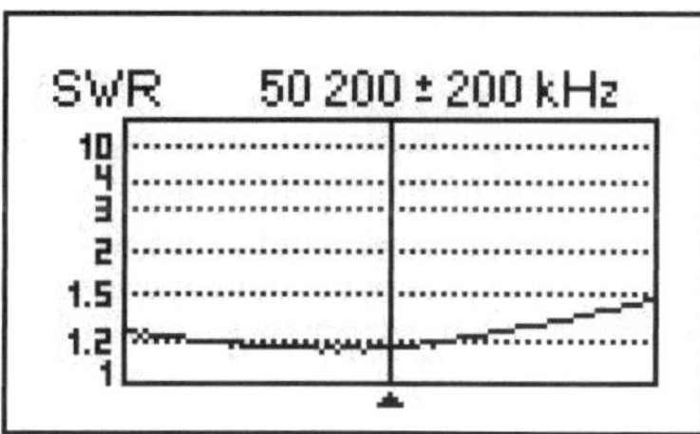

Figure 4.6 — The RigExpert AA-54 in swept frequency analysis mode. Note that in one view the operator can see the center tuned frequency and the bandwidth of the antenna system.

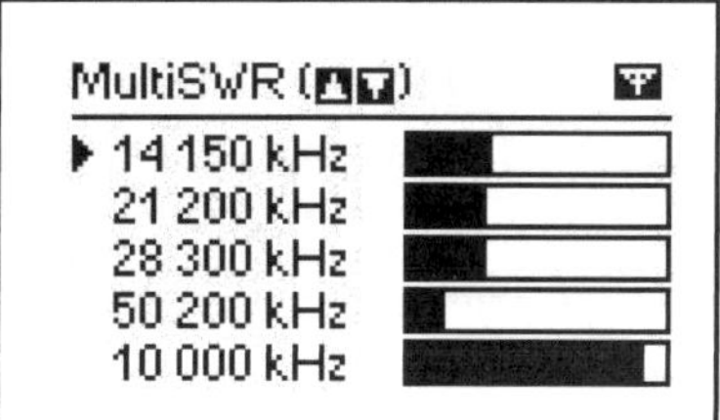

Figure 4.7 — The RigExpert AA-54 in multi-frequency SWR analysis mode. This is particularly useful for making adjustments on multi-band antennas in which a tuning adjustment on one band may affect a change to other bands.

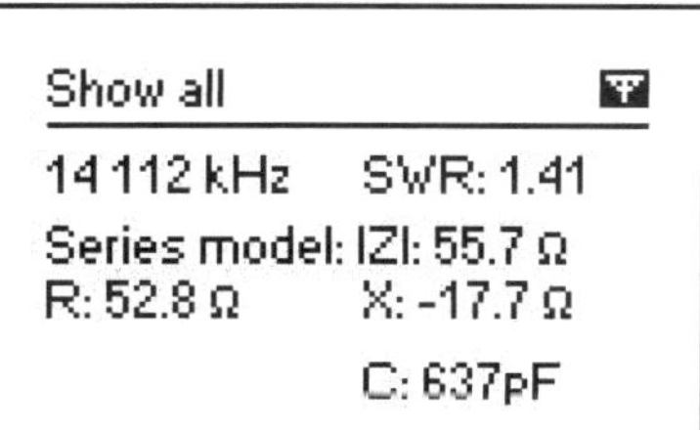

Figure 4.8 — The RigExpert AA-54 single frequency impedance display. Note that in addition to the SWR, it shows the magnitude of the impedance, the resistive and reactive part of the impedances, and even calculates the value of equivalent capacitance.

In addition to the single frequency display mode, several antenna analyzers offer swept frequency analysis capability. **Figure 4.6** shows the AA-54 in this mode. Note that in one view the operator can see the center tuned frequency and the bandwidth of the antenna system — a very useful arrangement. **Figure 4.7** shows the RigExpert AA-54 in a different kind of multi-frequency SWR analysis mode. In this case the SWR on up to five distinct frequencies is shown. This is particularly useful for making adjustments on multiband antennas in which a tuning adjustment on one band may affect tuning on other bands.

MEASURING IMPEDANCE

The impedance can be represented in several forms, as discussed in the previous chapter. **Figure 4.8**, again from a RigExpert AA-54 display, shows the whole story. Note that in addition to the SWR, it shows the magnitude of the impedance and the resistive and reactive part of the impedances. It even calculates the value of equivalent capacitance. Note that the series equivalent model is used, but an equally useful parallel component model can be specified, if desired.

In addition to single point numerical data, some analyzers offer graphical outputs over a frequency range. **Figure 4.9** shows a plot of impedance versus frequency as seen on the AA-54 handheld analyzer screen. Some analyzers can interoperate with a PC to provide data storage and display

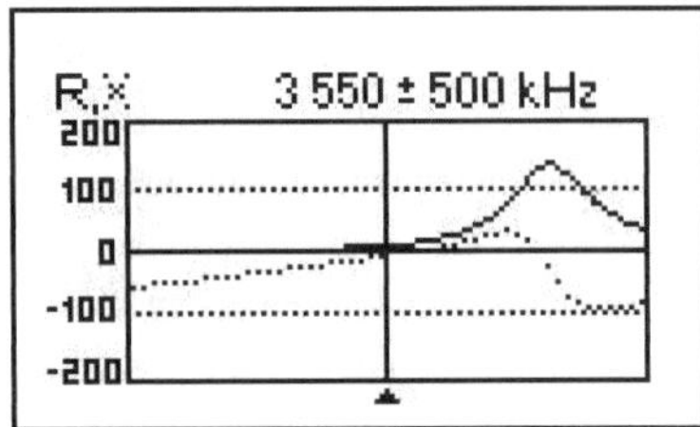

Figure 4.9 — Plot of impedance versus frequency as seen on the AA-54 handheld analyzer screen.

options not feasible on a compact handheld analyzer. The RigExpert analyzer family can work with supplied *AntScope* software to provide swept frequency impedance information (see **Figure 4.10**) with much more resolution than is available on the smaller handheld screen.

Note that the previous displays all depended on knowledge of the sign of reactance, or equivalently the sign of the phase angle of the impedance. While the AA-54 provides this directly, as do a few other units, most do not. Some have a sign, typically a plus sign, but a look at the manual indicates that this is a "place holder," not an indicator of the actual nature of the reactance.

Fortunately, in most cases it is pretty easy to tell whether the reactance is inductive or capacitive. This is because the reactance of an inductor goes up with frequency while that of a capacitor goes down. Thus if the frequency is increased *slightly* and the reactance goes up, we have an inductive (+) reactance; if it goes down it is capacitive (–). I say slightly because this only works if there is not a resonant point in between! On the other side of resonance, the sign reverses. Thus, if performing this test, make sure there is no change in direction of the change as you move between frequencies.

With the determination of the sign of the reactance, either explicitly or implicitly, as described, either of the preceding representations presents

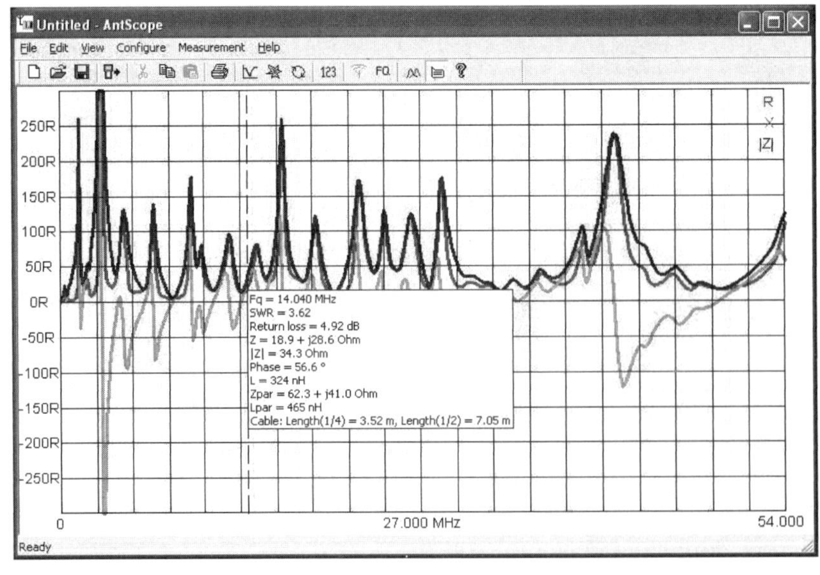

Figure 4.10 — Plot of impedance versus frequency as seen on a PC using RigExpert *AntScope* software with much more resolution than is available on the smaller handheld screen.

the whole impedance story, at least at the point of measurement. Some analyzers present the magnitude of the impedance without indicating the phase angle. This information, while useful in some applications, does not provide the complete story — for example a matching network cannot be easily designed based on the magnitude of the impedance, although it is possible to determine the impedance from this data with graphical or algebraic techniques, it is much more convenient to read it on the analyzer directly.[1]

[1]J. Stanley, K4ERO, "Determining Complex Impedance," *QST*, Sep 1996, p 40.

Chapter 5

Adjusting Your Antenna

The author adjusts the match to an antenna using an antenna tuner while observing the resulting SWR on the antenna analyzer.

USING YOUR ANTENNA ANALYZER DATA TO TRIM ANTENNAS

Most antennas are designed to operate over a range of frequencies and are specified to exhibit a particular impedance at each frequency. Often, rather than specifying the actual impedance, the designer or manufacturer will specify a nominal impedance and SWR over a range of frequencies.

This sounds quite straightforward, and would be were it not for the fine print. To be complete, the designer should also specify the design height above ground, the electrical characteristics of the ground, and how far the antenna needs to be from other objects that it could interact with. As a practical matter, even if you knew all those parameters, you would likely be stuck with putting your antenna in the available space over your real ground at whatever height you have supports for.

A Typical Example

The magnitude of the effect of the differences in conditions from those of the ideal design environment will depend on a number of factors, but we can get a feel for the amount of change that might be encountered by looking at a simple example. We will select a wire 40 meter dipole as shown in **Figure 5.1**. The example we will use is made of bare #14 AWG wire, tuned to be resonant at 7.15 MHz in free space, resulting in a total length

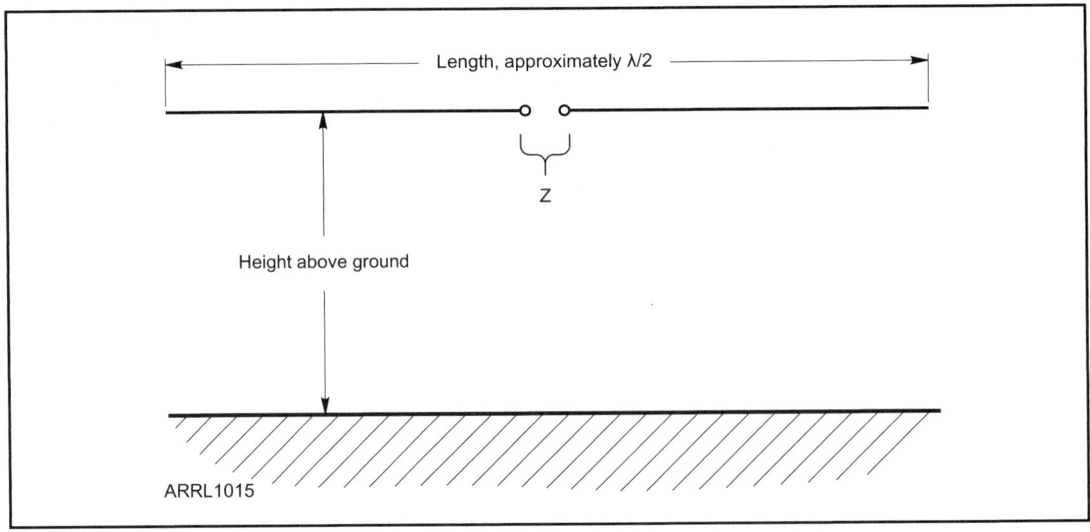

Figure 5.1 — The configuration of a 40 meter dipole, our representative sample.

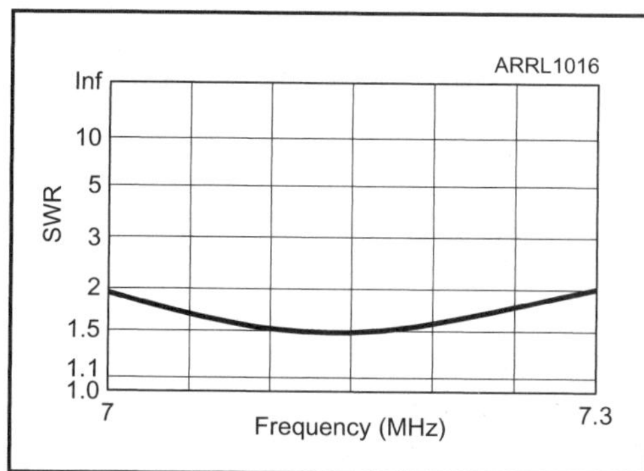

Figure 5.2 — SWR response of a 40 meter dipole in free space cut for resonance at 7.15 MHz.

of 66.9 feet. This will be our point of comparison. The SWR of such a dipole is shown in **Figure 5.2**.

Height Above Ground and Ground Characteristics

The height above ground of a horizontal dipole antenna has a major impact on a number of antenna operational characteristics. The vertical (and even the horizontal to a lesser extent) radiation patterns are a function of height. Usually more noticeable, however, is the resonant frequency and impedance at resonance as shown in **Table 5.1**. This table shows the *EZNEC* antenna modeling software prediction of the resonant impedance of a dipole at different heights above both perfect and typical ground.[1] Your ground is probably closer to "typical" (see the note in Table 5.1) than to free space, although the higher the antenna is placed, the closer its characteristics are to a free space antenna.

Because the characteristics of an antenna are dependent on factors beyond the control of the designer, most antennas provide an adjustment mechanism to allow compensation after installation in the environment where the antenna will be used.

Table 5.1
Variation in Dipole Resonant Frequency and Impedance with Height Above Ground*

	-------Typical Ground -------		*----- Perfect Ground-----*	
Height (λ)	*Frequency (MHz)*	*Impedance (Ω)*	*Frequency (MHz)*	*Impedance (Ω)*
Free Space	7.15	73.5		
1	7.175	72	7.19	72
0.75	7.1	75	7.18	60
0.5	7.2	68	7.23	70
0.4	7.2	84	7.21	92
0.3	7.13	89	7.09	95
0.2	7.05	73	7.00	63
0.1	7.08	53	7.07	22

*"Typical ground" parameters; relative dielectric constant 13, conductivity 0.005 S/m. Bare #14 AWG copper wire.

Making the Adjustments

If we were to install our nominal half-wave dipole 0.2 wavelength above real ground, we would find that the measured resonant frequency is at 7.05 MHz, rather than the design value of 7.15 MHz. See **Figure 5.3**. While the difference may not sound great, the more significant effect is that the original design had a 2:1 SWR covering the whole 40 meter amateur band, while at this height SWR now is above 2:1 at frequencies above 7.2 MHz.

With an antenna that is resonant at a frequency below the desired frequency, the rule is to shorten the antenna until the resonance moves where we want it. Some amateurs do it a bit at a time, watching the frequency rise until they have gone too far and then backing up a bit. While this can work, it can be tedious because each step requires the antenna be lowered, adjusted and then raised again. A more efficient approach is to recognize that the required length change is inversely proportional to the required frequency change. Since we want to shorten the antenna in this case, the new length should be 7.05/7.15 × 66.9 or 65.96 feet total — remove 5.6 inches from each end. Rather than cutting the wire, the excess can be folded back along the antenna until you are satisfied with the result.

With an antenna that is not resonant at the desired frequency, it is a bit more complicated. An example is the *extended double Zepp* (EDZ) antenna. An EDZ is configured in the same manner as a half wave dipole, except it is longer — its nominal length is 1.25 wavelengths. This antenna has a broadside gain of about 3 dB with a corresponding narrow azimuth pattern and is usually fed with a matching network or a matching section of mismatched transmission line.

An EDZ made of the same wire as our dipole would be about 167 feet long in free space and would have an impedance at the 7.15 MHz center frequency of 224 – *j*1138 Ω. This is a bit of a problem to measure directly, since that represents a 50 Ω SWR of 117:1, putting it out of the measurement range of most antenna analyzers. All is not lost. We can use a few different approaches to make sure it is the correct length when installed over real ground.

1) Don't worry about it. Since it doesn't need to be resonant, we can just build the antenna as designed and install it and hope for the best. Small changes in the length won't make a big difference in the pattern or in the impedance, and can likely be accommodated if the matching network is adjustable. While this is reasonable

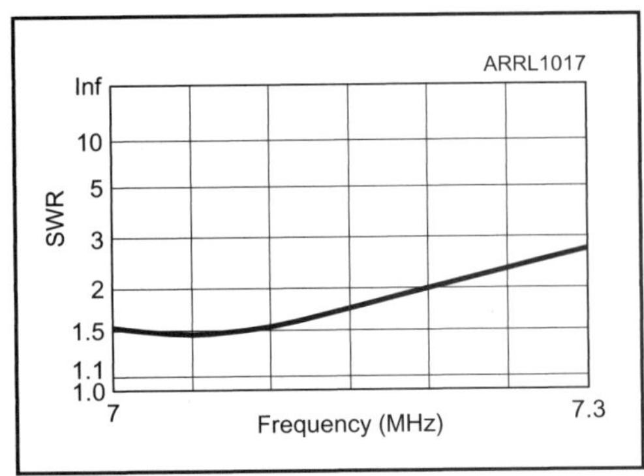

Figure 5.3 — SWR response of the 40 meter dipole from Figures 5.1 and 5.2 installed at 0.2 wavelength above real ground.

with the EDZ, it may not be viable for all nonresonant antennas.

2) Test the antenna and matching network together. If the matching network is designed to match the antenna to 50 Ω, by connecting the antenna to the matching network we can adjust the antenna length until we have a 1:1 SWR at the network input. This is a fine approach, but only if we know the matching network or section is trimmed properly — not usually the case with a new setup.

3) Make use of the antenna's resonant frequency. While we aren't using the EDZ at its resonant frequency, it is a resonant half wave dipole at some other frequency. We can measure and adjust to that frequency. Using modeling software, we find the half wave resonant frequency of a 7.15 MHz EDZ is 2.87 MHz. Thus if we trim the antenna to be resonant in its surroundings at that frequency, it will be just the right length to be a 40 meter EDZ. In this way we can optimize the antenna first and then use the analyzer to tune the matching network section to have an input Z of 50 Ω.

TUNING AT THE STATION END OF THE TRANSMISSION LINE

In some stations, the final antenna adjustments are accomplished at the station end of the transmission line using an antenna tuner, which is an adjustable impedance matching device.[2] Tuning at the transmitter end does not change the SWR on the transmission line, but can make the transmitter "see" a matched load. This can be a useful technique, especially for use with antennas that are operated outside their normal SWR bandwidth.

If the antenna tuner is a manually adjusted one, it will require a signal on the desired frequency in order to view the results of the adjustments. Many operators use the station transmitter for this purpose, but there are two issues with this approach.

Interference. While the tuner is being adjusted, the transmitter is on the air, often for lengthy periods as the optimum settings are found. During this time, your signal will interfere with others using the frequency. Even listening on the channel is no guarantee of avoiding this because the station you are interfering with may not be one you can hear, even though the receiving station can hear you as well as the other station in the QSO.

Stress on components. Until the tuner is adjusted, the transmitter is driving into a mismatched load. While most modern transmitters include circuitry to reduce power automatically to avoid damage, not all do, and those that do may not be working properly. Of course it is always a good idea to reduce power manually while tuning, but not all transmitters are equipped to do that easily.

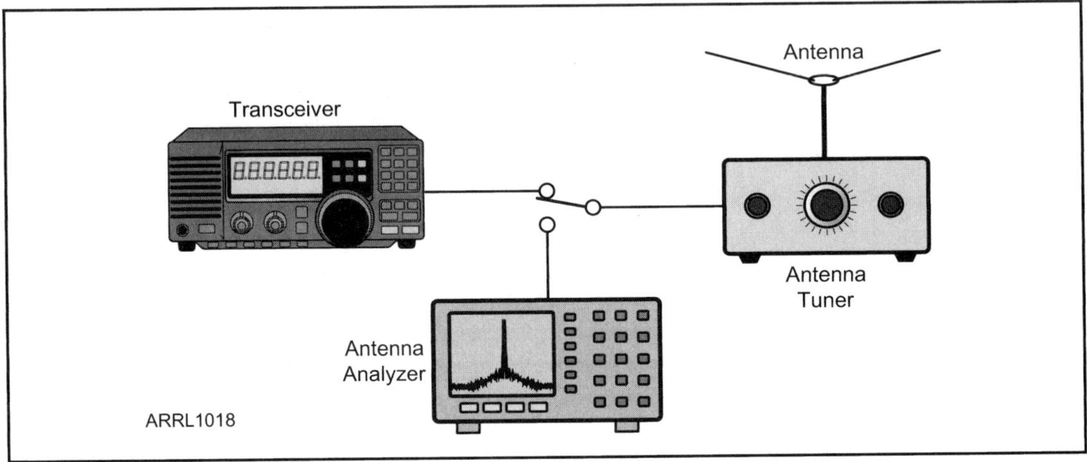

Figure 5.4 — By switching the transmitter end of an antenna tuner to an antenna analyzer, adjustments can be made without causing harmful interference to other users. This also avoids unneeded stress on the tuner components while looking for a match.

The solution is to use the antenna analyzer to adjust the antenna tuner. Just switch the tuner input from the radio to the analyzer, set the analyzer to the desired frequency and adjust the tuner for minimum SWR (see **Figure 5.4**).

OBSERVING CHANGE AS MEASUREMENT CONDITIONS ARE VARIED

The previous section, describing the initial adjustment of antennas, is important and may be the primary reason amateurs are willing to invest in an antenna analyzer. If the work stops there, many opportunities will be lost. An antenna analyzer can provide an opportunity to track changes to the system after it is installed.

Record Your Data

It is very important that the antenna measurement data be recorded upon completion of tuning. In this way any changes caused by component failure or degradation can be noted. Be particularly wary of improved SWR over time. Since antennas rarely improve over time, it is very likely that improved SWR or SWR bandwidth is an indication of additional loss, often in a transmission line or related component or perhaps caused by corroded connections.

Antenna Changes During Normal Operation

Once an antenna is installed with appropriate compensation for its surroundings, it is important to keep track of what is going on around it. A rotating directional array, for example, may be properly tuned while looking toward the open sky, but may have a significant impedance difference if pointed into a neighboring building. It may be possible to find a set of adjustments that provides reasonable performance at all azimuths, or in the worst case, operation may need to be avoided or power reduced with the antenna pointed in certain directions.

A change in measured data may also indicate a change in surroundings. Construction work of various forms can result in new, perhaps hidden, metal structures that could change antenna characteristics. As the forest marches on, it also gains in height. Antennas, particularly at VHF and UHF, suffer if the former shot to a clear sky becomes crowded by trees that are suddenly higher than the antenna.

Notes

[1]Several versions of *EZNEC* antenna modeling software are available from developer Roy Lewallen, W7EL, at **www.eznec.com**.

[2]J. Hallas, W1ZR, *The ARRL Guide to Antenna Tuners*. Available from your ARRL dealer or the ARRL Bookstore, ARRL order no. 0984. Telephone 860-594-0355, or toll-free in the US 888-277-5289; **www.arrl.org/shop**; **pubsales@arrl.org**.

Chapter 6

Taking the Feed Line into Account

Coaxial cables going up one of the towers at W1AW, the ARRL Headquarters station.

Mismatched transmission lines transform impedances based on the degree of mismatch and electrical length. Thus the impedance and resonant frequency measured at the transmitter end of the feed line may be very different from that at the antenna load. Unless this is taken into account, the measurement results will be misleading.

USING SWR TO DETERMINE REMOTE ANTENNA CHARACTERISTICS

In most real world cases, it is both impractical and undesirable to measure the impedance or SWR of an antenna at the actual antenna terminals. For one thing, the antenna terminals may be suspended in mid air, as in the dipole shown in the previous chapter. If we lower the antenna to be within reach, the impedance and resonance of the antenna will change as discussed there. Another problem is that if the SWR meter and operator are located at the antenna terminals, in almost all cases they will be within the antenna's near field and will affect the readings.

While the impedance will be transformed over distance by a length of mismatched transmission line, the SWR data will still be usable. In fact, if the antenna is resonant and matched to the transmission line Z_0 at a particular frequency, by definition its impedance will be equal to that of the line Z_0, and it will be the same value at the radio end of the line. While this sounds like a special case, it is actually one that we frequently will be trying to obtain.

Note that the above relationship holds whether the resonant frequency is the one we are looking for or not. If we want to adjust an antenna to a resonant frequency, and the antenna is connected to the analyzer through a transmission line, it is best to find the antenna's resonant frequency rather than the impedance at the desired frequency. Knowing the resonant frequency of the antenna will tell us what we need to know in order to adjust the antenna for the desired frequency. This approach is illustrated in **Figures 6.1** through **6.4**.

In Figure 6.1, we see the SWR plot of a 66.4 foot long, low, wire, 40 meter dipole resonant at 7.15 MHz, as measured (modeled) at the antenna terminals. The height was adjusted to obtain the nice match to 50 Ω. Impedance is 52.2 – j0.05 Ω. Since we probably won't be using a tall ladder to measure at the feed point, we have Figure 6.2. This is the SWR plot of a the same 40 meter dipole as in Figure 6.1, but as measured at the end of a 45 foot, ideal, lossless 50 Ω transmission line. Impedance is 48.8 + j1.7 Ω. Note that while the SWR is virtually the same, the reactive part of impedance has changed sign due to the transformation in the line. By looking at this plot, we really know everything we need to in terms of how our transmitter will operate with this antenna system, except we can't tell (and don't often care) what the actual impedance is at the antenna.

If upon installation our antenna is not resonant at the desired frequency, but needs adjustment, we might see an SWR plot such as in Figure 6.3. This is the SWR plot at the antenna terminals of a low 67.2 foot long, 40 meter dipole before being adjusted to resonance at 7.15 MHz, our desired frequency. Impedance is 54.1 + j22.3 Ω. Note that if measured at the antenna terminals, the impedance at 7.15 MHz shows a positive reactance, indicating

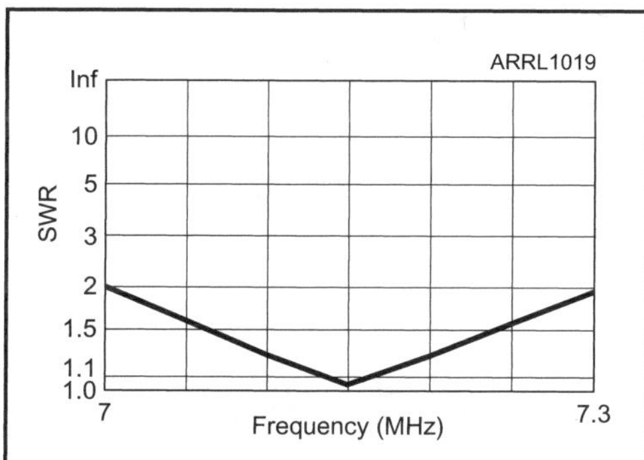

Figure 6.1 — SWR plot of a low 40 meter dipole of length 66.4 feet and resonant at 7.15 MHz, as measured (modeled) at the antenna terminals. The modeled impedance at 7.15 MHz was 52.16 –*j*0.47 Ω.

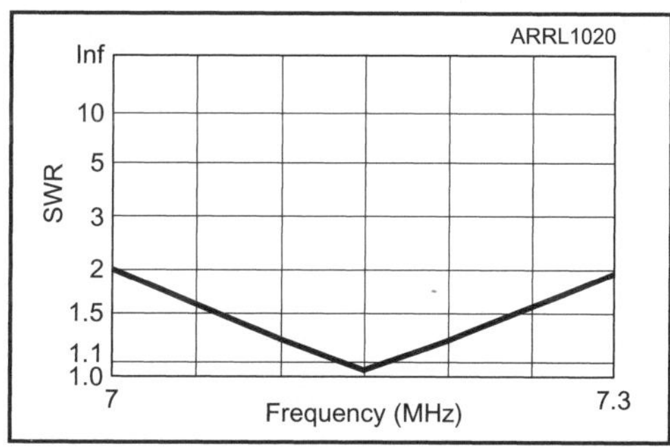

Figure 6.2 — SWR plot of a the same 40 meter dipole as in Figure 6.1, but measured at the end of 45 feet of ideal, lossless 50 Ω transmission line. Note that while the SWR is virtually the same, the reactive part of impedance has changed sign due to the transformation in the line as described in the text. The modeled impedance at 7.15 MHz was 48.83 +*j*1.683 Ω.

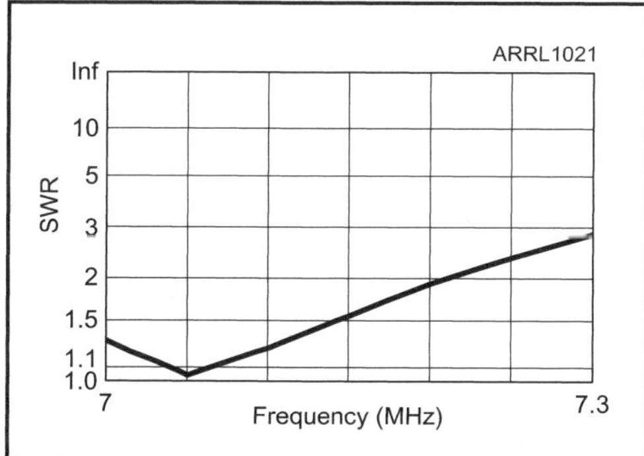

Figure 6.3 — SWR plot at the antenna terminals of a low 67.2 foot 40 meter dipole before being adjusted to resonance at 7.15 MHz. The modeled impedance at 7.15 MHz was 54.15 +*j*22.28 Ω.

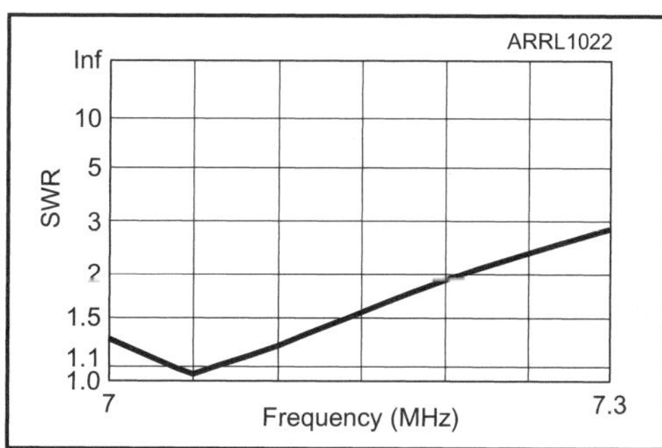

Figure 6.4 — SWR plot of a the same 40 meter dipole as in Figure 6.3, but measured at the end of 45 feet of ideal, lossless 50 Ω transmission line. Note that while the 7.15 MHz SWR is virtually the same, the reactive part of impedance has changed sign due to the transformation by the mismatched transmission line. The modeled impedance at 7.15 MHz was 32.65 –*j*3.007 Ω.

that the antenna is inductive or too long. This is confirmed by the fact that the resonant frequency is lower than desired.

Figure 6.4 is the SWR plot of the same 40 meter dipole as in Figure 6.3, but measured at the bottom end of 45 feet of ideal, lossless 50 Ω transmission line. Impedance is $32.6 - j3.0$ Ω. Note that while the 7.15 MHz SWR is virtually the same, the reactive part of the impedance has changed sign due to the transformation by the mismatched transmission line. If we just looked at the impedance at 7.15 MHz, we would note that at the bottom end of the line it is capacitive. From this single data point, we might wrongly conclude that the antenna was too short to be resonant at our desired frequency.

This illustrates the benefit of watching the SWR if adjusting antennas that have an SWR of close to 1:1 at *some* frequency. If we look at the frequency at which the SWR is best (in this case 7.05 MHz), instead of looking at the impedance at our desired frequency, we can conclude that the antenna resonant frequency is too low and the antenna needs to be shortened.

To make the adjustments, you could trim an inch at a time from each end until the frequency of lowest SWR coincides with the desired frequency. While this approach works, it can be a bit tedious. In addition, using this approach the resonant frequency often ends up at least slightly above the target frequency. I recommend moving in more quickly by noting that the ratio of the desired frequency to measured frequency is very close to the inverse of the ratio of the lengths. Thus, if the measured frequency, F_M, is 7.05 MHz, and the desired frequency, F_D, is 7.15 MHz, the new length will be very close to the current length (67.2 feet) times 7.05/7.15 = 66.3 feet, quite close to the 66.4 foot length of the antenna tuned at 7.15 MHz.

Antenna builders take note — while trimming antennas, I suggest you fold the wire back on itself at the end insulators rather than cutting it off. This allows adjustments to be undone if required and also provides some extra in case the wire breaks at the connection end in the future and the antenna needs to be reinstalled.

TRANSLATING IMPEDANCE AT THE MEASUREMENT END TO THE ANTENNA END

The preceding discussion of using SWR as a technique for adjusting antennas was predicated on the antenna system being matched at some frequency. While often the case, it is not always true. Sometimes it is important to know what the actual impedance is at the antenna following measurement at the end of a connected transmission line. Fortunately, there are a number of ways to accomplish this, as we will describe below. Each method requires that we know the characteristics of the transmission line, including its physical length, or in some cases, electrical length.

Transformation Using Transmission Line Equations

The complex impedance at the bottom of a transmission line (Z_X) of length x and characteristic impedance Z_0 terminated with a complex load of impedance Z_L can be found from the following equation:[1]

$$Z_X = Z_0 \times \frac{Z_L + Z_0 \tanh \gamma x}{Z_0 + Z_L \tanh \gamma x} \qquad \text{(Eq 6.1)}$$

where tanh is the hyperbolic tangent function, and $\gamma = \alpha + j\beta$ (α is the attenuation constant in nepers/meter and $\beta = 2\,\pi/\lambda$, the phase constant in radians/meter, where λ is the wavelength in meters).

If the cable is lossless, a reasonable assumption for short transmission lines at HF, Eq 6.1 simplifies to:

$$Z_X = Z_0 \times \frac{Z_L + jZ_0 \tan \beta x}{Z_0 + jZ_L \tan \beta x} \qquad \text{(Eq 6.2)}$$

While one or both of these can be set up on a spreadsheet or programmable calculator, note that for this purpose we would really like to obtain Z_L based on a measurement of Z_X in order to know the impedance at the antenna based on the measured impedance at the bottom of the line. I will leave that exercise to the zealous reader more adept than I am with manipulation of complex algebraic functions — especially since I will describe some easier methods.

Transformation Using Graphical Techniques

Long before the advent of programmable calculators and personal computers, graphical techniques were a popular solution method for otherwise tedious calculations. The *Smith Chart* is such a device, invented before the Second World War by Phillip H. Smith.[2] Smith Charts (see **Figure 6.5**) are available in both normalized version (center is 1.0 as in Figure 6.5) or preset for a specific impedance (usually 50 Ω).[3] To use the normalized chart, divide the real and imaginary parts of the impedance by the transmission line Z_0 and enter the value as a point on the chart. In this example an impedance of 35 + *j*12.5 Ω is normalized to the line Z_0 of 50 Ω as 0.7 + *j*0.25 Ω by dividing each term by the Z_0 and entered as shown. If a 50 Ω chart is used, the real and imaginary results can be entered directly.

A circle with its center at the 1.0 + *j*0 point drawn through the entered point is a circle of constant SWR with the SWR value indicated by the intersection of the circle with horizontal axis on the right side of the center, 1.6:1 in this example.

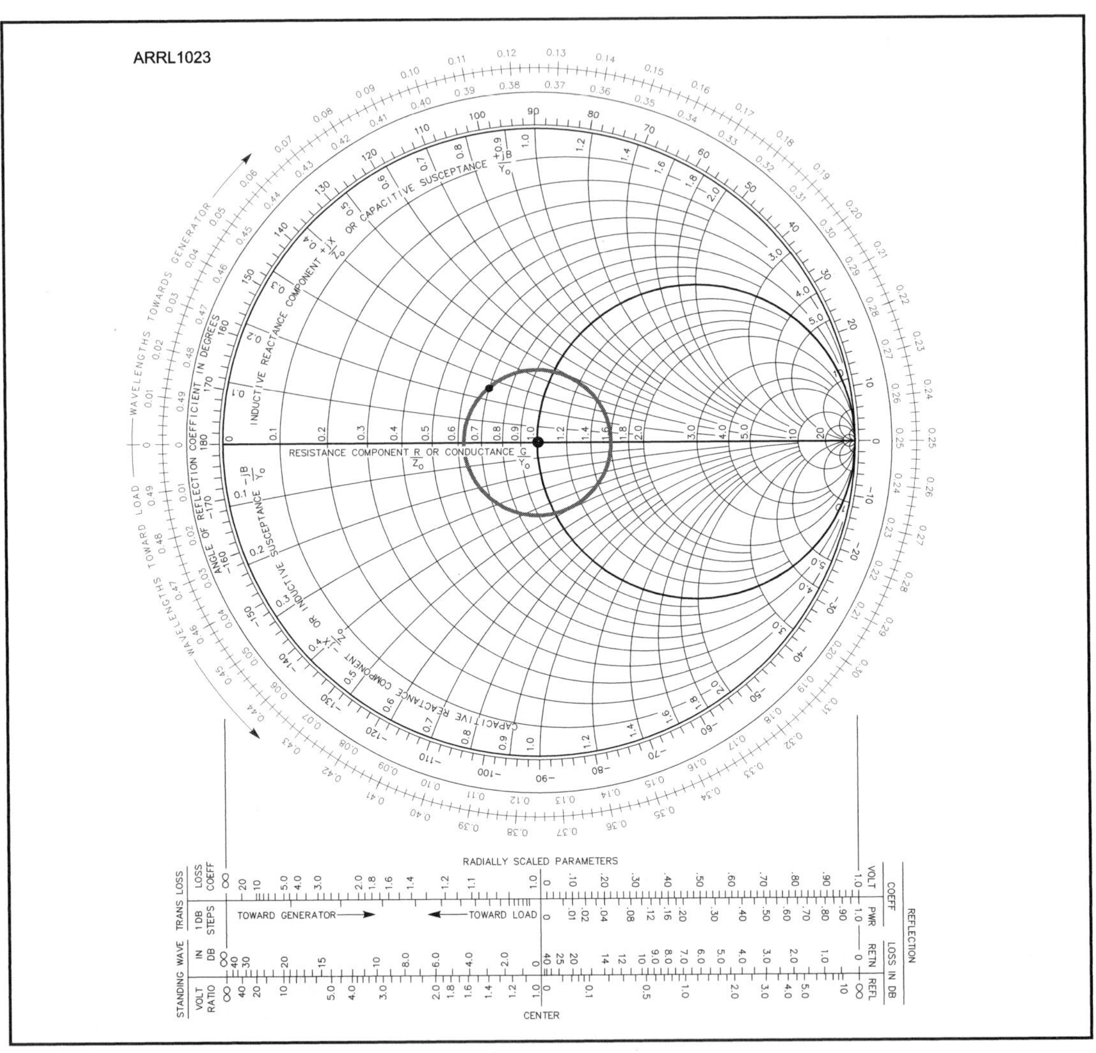

Figure 6.5 — A normalized Smith Chart. To use, divide the real and imaginary parts of the impedance by the transmission line Z_0 and enter the value as a point on the chart. In this example an impedance of 35 + *j*12.5 Ω is normalized to the line Z_0 of 50 Ω as 0.7 + *j*0.25 Ω and entered as shown.

To find the answer we were looking for, we first draw a line from the center through the entered point on the chart and extend it to the outer edge of the chart as shown in **Figure 6.6**. If the impedance were that measured at the bottom of our transmission line, we would find a point on the circumference going counterclockwise (*toward load* as indicated) moving a distance equal to the electrical length of the line. Where the line to the new point intersects

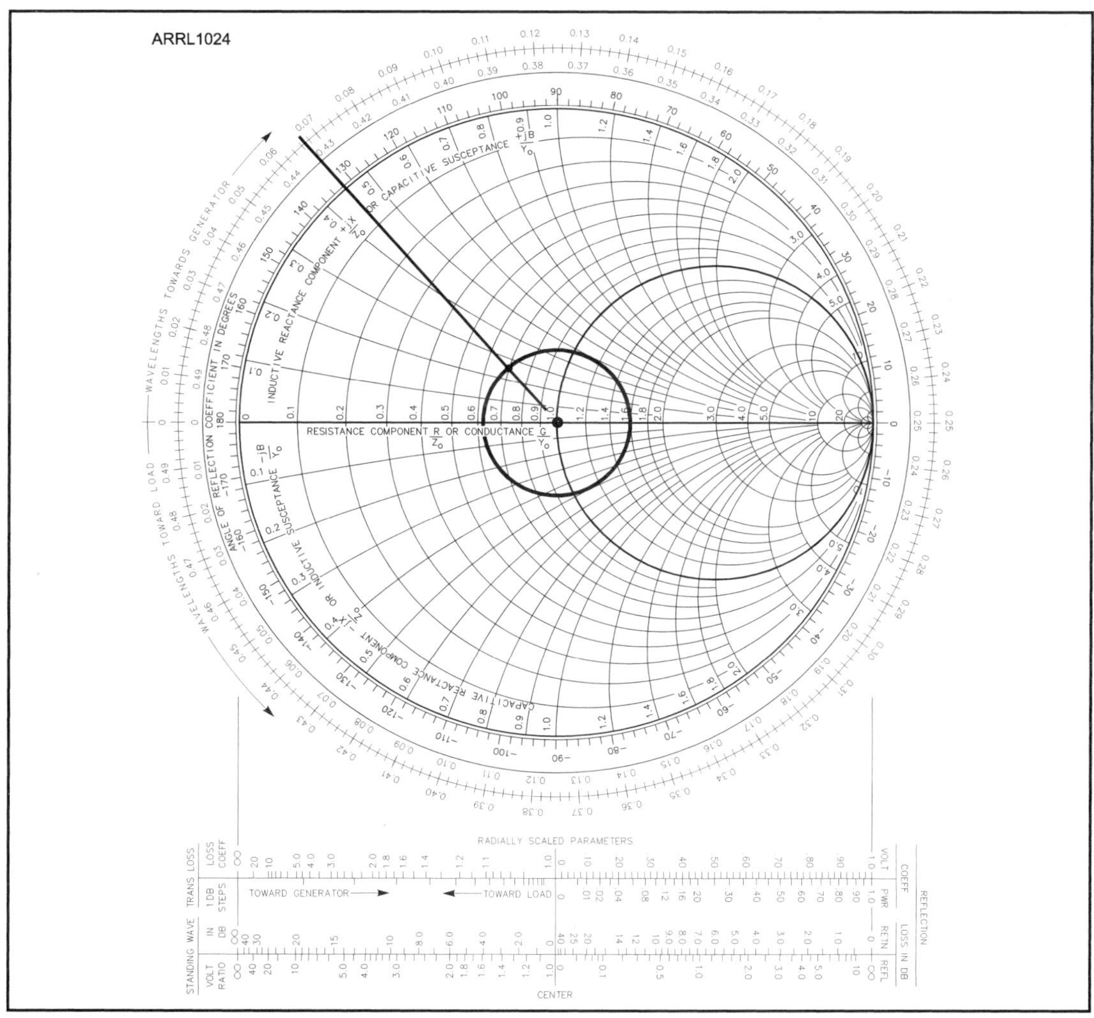

Figure 6.6 — A normalized Smith Chart used to calculate antenna impedance from impedance measured at the end of the line. The line shown from the center to the constant SWR circle is rotated "toward the load" the number of degrees corresponding to the electrical length of the line. The point at which it intersects the constant SWR circle represents the impedance at the antenna.

the constant SWR circle, we read our antenna impedance. Note that for lines longer than ½ wavelength (180°), we just go around and around as needed. For a normalized chart, we multiply the resulting impedance values by the Z_0.

This method is as viable today as it was 70 years ago, however it does have the limitation that it assumes a lossless line. As noted previously, this is not a serious source of error in many instances.

Transformation Using Transmission Line Software

Packaged with the last few editions of *The ARRL Antenna Book* is a CD with some very useful software. The program that I use almost daily is one written by retired *ARRL Antenna Book* Editor R. Dean Straw, N6BV, called *TLW* for *Transmission Line for Windows.*

TLW provides a very easy to operate mechanism to determine everything I usually need to know about what's happening on a transmission line. When you open the program, you are presented with a screen as shown in **Figure 6.7**. This has the values plugged in from the last time you used it, often saving a step. Let's take a quick tour of the inputs:

- *Cable type* — allows you to select the cable you would like to analyze. A drop-down box provides for the selection of one of 32 of the most common types of coax and balanced lines. An additional entry is provided for *User Defined Transmission Lines* that can be specified by propagation velocity and attenuation.
- *Length* — in feet or meters, your choice.
- *Frequency* — an important parameter when dealing with transmission line effects.
- *Source* — defines the form of the input impedance. Generally, you can use NORMAL.
- *Impedance* — can be specified as what you measure, resistive (real) and reactive (imaginary, minus means capacitive). This could come from your antenna analyzer at either end of the transmission line. Note: If you know only the SWR, not the actual impedance, all is not lost — see below.

The available outputs include:

- *SWR* — is provided at each end of the cable. This is an important difference that many people miss, important even with a moderate SWR at the transmitter end. As we'll see, the SWR at the antenna can be much higher because of the cable loss. With *TLW*, you know instantly the SWR at both ends,

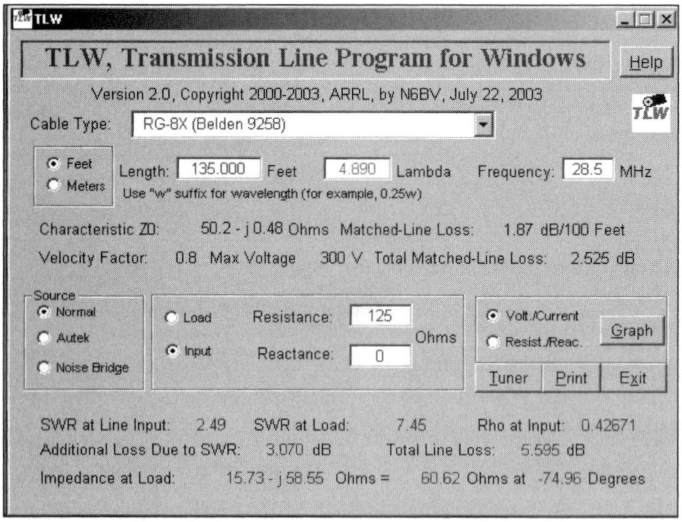

Figure 6.7 — The opening screen of *TLW*, illustrating the process described in the text. The results are interesting. Note that the 2.5:1 SWR as seen at the radio on 28.5 MHz through 135 feet of RG-8X results from a 7.45:1 SWR at the antenna.

as well as the loss in the cable itself

• *Rho at input* — the reflection coefficient, the fraction of the power reflected back from the load.

• *Additional loss due to SWR* — one of the answers we were after.

• *Total line loss* — is the other, the total loss in the line, including that caused by the mismatch.

Often the only measurement data available is the SWR at the transmitter end of the cable, rather than the complex impedance — depending on the analyzer capability. Because the losses are a function of the SWR, not the particular impedance, you can just enter an arbitrary impedance with that same SWR and click the INPUT button. An easy arbitrary impedance to use is just a resistance value of the SWR times the Z_0 of the cable, usually 50 Ω. For example, you could use a resistance of 125 Ω to represent an SWR of 2.5:1. This is what we've done in **Figure 6.8**, using 135 feet of popular Belden RG-8X.

The results are interesting. Note that the 2.5:1 SWR as seen at the radio on 28.5 MHz results from a 7.45:1 SWR at the antenna — perhaps this is an eye-opener! Note that of the 5.6 dB loss, more than half, or 3.1 dB, is due to the mismatch. If we used something other than the actual measured impedance, we can't make use of the impedance data that *TLW* provides. We can use the SWR and the resulting loss data, but that's probably what we wanted to find out.

We can now do some "what ifs." We can see how much loss we have on other bands by just changing the frequency. For example, on 80 meters, with the same 2.5:1 at the transmitter end, the SWR at the antenna is about 3:1 and the loss is slightly more than 1 dB. We could also plug in an impedance calculated at the antenna end and see what difference other cable types would make. For example, with the same 28.5 MHz SWR of 7.45 at the antenna and 135 feet of ½ inch Andrew Heliax, we will have a total loss of 1.5 dB at 28.5 MHz. Note that the SWR seen at the bottom will now be 5.5:1 and our radio's auto-tuner

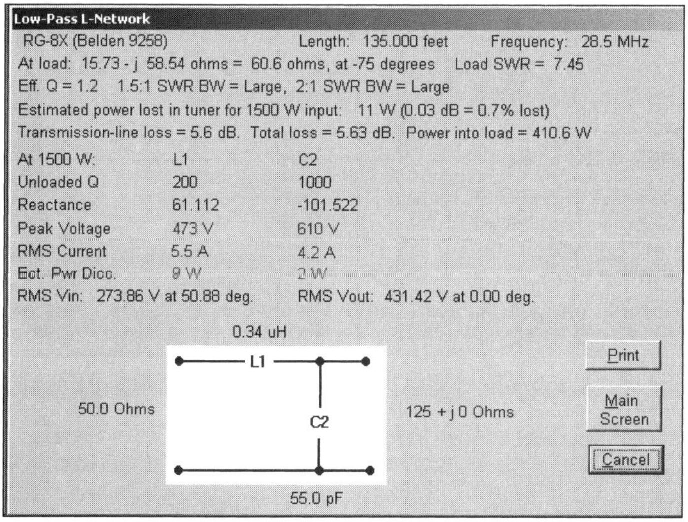

Low-Pass L-Network

RG-8X (Belden 9258)   Length: 135.000 feet   Frequency: 28.5 MHz

At load: 15.73 - j 58.54 ohms = 60.6 ohms, at -75 degrees   Load SWR = 7.45

Eff. Q = 1.2   1.5:1 SWR BW = Large, 2:1 SWR BW = Large

Estimated power lost in tuner for 1500 W input: 11 W (0.03 dB = 0.7% lost)

Transmission-line loss = 5.6 dB. Total loss = 5.63 dB. Power into load = 410.6 W

At 1500 W:	L1	C2
Unloaded Q	200	1000
Reactance	61.112	-101.522
Peak Voltage	473 V	610 V
RMS Current	5.5 A	4.2 A
Est. Pwr Diss.	9 W	2 W

RMS Vin: 273.86 V at 50.88 deg.   RMS Vout: 431.42 V at 0.00 deg.

0.34 uH
L1
50.0 Ohms
C2
125 + j 0 Ohms
55.0 pF
Print
Main Screen
Cancel

Figure 6.8 — The antenna tuner screen of *TLW*. Additional antenna tuner losses are described.

might not be able to match the new load — an unintended consequence of making the system "better."

You can also click the GRAPH button and get a plot of either voltage and current or resistance and reactance along the cable. These will only be useful if we have started with actual impedance, rather than SWR.

Pushing the TUNER button results in a page asking you to select some specifications for your tuner parts and then provides the results shown in the lower section of Figure 6.8. As shown, *TLW* effectively designs a tuner of the type you asked for at the shack end of the cable. It indicates the component voltages and currents — important to specify the parts needed. It also calculates the power lost in the tuner and gives a summary of the transmitted and lost power in watts, so you don't need to calculate it!

When you've finished, be sure to hit the EXIT button, don't just close the window. Otherwise *TLW* may not start properly the next time you want to use it.

Measuring with a ½ Wavelength Transmission Line

A lossless electrical ½ wavelength transmission line has a sometimes useful property: It will repeat the impedance that is present at the far end at the near end. This means that if have an antenna connected to our analyzer by a lossless line of an electrical ½ wavelength, the impedance we measure at the transmitter end will be exactly the same as the impedance we would measure at the antenna, independent of the transmission line Z_0. In fact this holds true for any integral multiple of half waves. The next chapter includes a section on adjusting the electrical length of a transmission line using your antenna analyzer, so this is a feasible approach.

Table 6.1
Complex Impedance Measured at Antenna Compared to Measurement at End of ½ and 5⁄2 Wave 450 Ω Transmission Lines Resonant at 7.15 MHz.

	At Antenna		*½ Wave 450 Ω Line*		*5⁄2 Wave 450 Ω Line*	
Frequency (MHz)	*R (Ω)*	*jX (Ω)*	*R (Ω)*	*jX (Ω)*	*R (Ω)*	*jX (Ω)*
7.00	84.1	–33.3	87.7	–59.0	109.5	–167.4
7.05	85.9	–22.2	88.3	–39.2	102.0	–108.3
7.10	87.6	–11.1	89.6	–19.6	98.7	–53.2
7.15	89.4	0.0	91.3	0.0	93.6	0.0
7.20	91.2	11.0	93.3	19.4	102.4	52.9
7.25	93.0	22.1	95.6	39.0	109.9	107.4
7.30	94.9	33.1	98.3	58.5	122.6	165.0

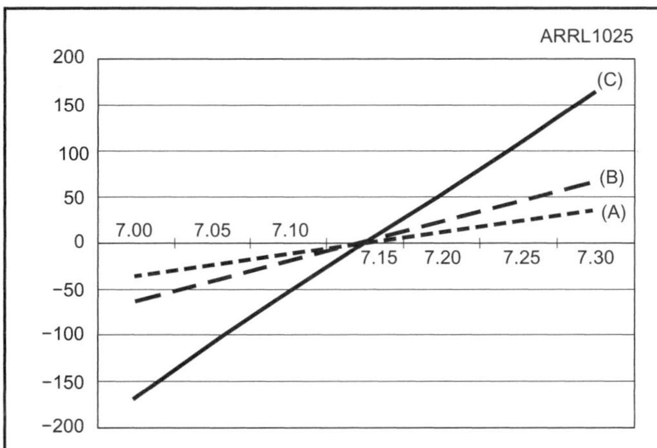

Figure 6.9 — Reactance of a 45 foot high 40 meter dipole resonant at 7.15 MHz measured at the antenna (A) at the end of a real 450 Ω transmission line with an electrical length of ½ wavelength (B) and 5⁄2 wavelength (C) at 7.15 MHz.

There are a number of caveats, however. First, this only gives accurate results if the line is indeed lossless (the higher the loss, the lower the accuracy). Similarly, it provides only an accurate measurement at the frequency at which the line is a multiple of ½ wavelength. Thus, sweeping over a band looking for the frequency of minimum SWR will give you the minimum SWR, but the impedance will only be that of the antenna at the ½ wavelength frequency.

Table 6.1 and **Figure 6.9** illustrate the point. Note that while the impedance at the resonant frequency is quite close for all cases, the reactance is quite different as the frequency moves from resonance. Still, the data would assist in trimming an antenna to resonance. Note also that multiple half wavelengths (the 5⁄2 wavelength column) multiply the apparent error. The difference in the resonant impedance can be attributed to the losses in the transmission line acting as additional resistance.

Notes

[1]J. Kraus, W8JK (SK), *Electromagnetics*, Sec. 11-6. McGraw Hill, New York, 1953.

[2]P. Smith, "Transmission Line Calculator," *Electronics*, Vol. 12, No. 1, Jan 1939, pp 29-31.

[3]Regular (ARRL order no. 1340) and expanded scale (ARRL order no. 1350) Smith Charts are available from your ARRL dealer or the ARRL Bookstore. Telephone 860-594-0355, or toll-free in the US 888-277-5289; **www.arrl.org/shop/**; **pubsales@arrl.org**.

[4]H. Ward Silver, NØAX, Ed., *The ARRL Antenna Book*, 22nd Edition, available from your ARRL dealer or the ARRL Bookstore, ARRL order no 9043. Telephone 860-594-0355, or toll-free in the US 888-277-5289; **www.arrl.org/shop/**; **pubsales@arrl.org**.

Chapter 7

Other Antenna Analyzer Applications

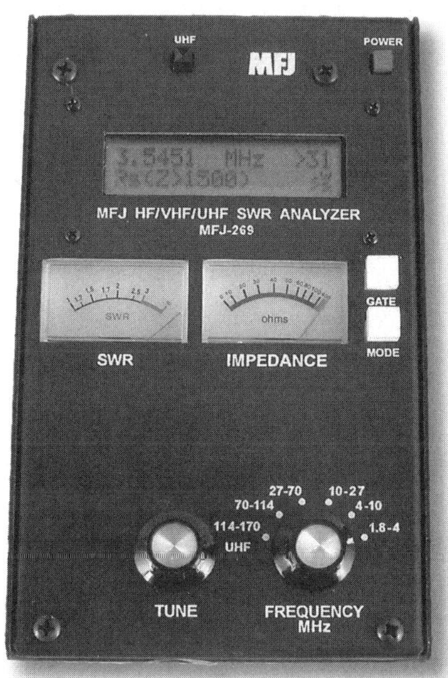

The MFJ-269 provides a digital signal source as well as a direct input to the internal frequency counter. While these features combine to make a basic general purpose instrument, other analyzer models may meet specific needs as well.

As noted in Chapter 2, antenna analyzers include multiple subsystems to generate and measure RF signals. In many cases, these subsystems can be used for other functions in addition to antenna analysis. The additional capabilities depend on the details of the design and the features built in. Not every analyzer can perform each function described below, but all analyzers can do some of them.

USING AN ANTENNA ANALYZER AS A TEST INSTRUMENT

Antenna analyzers can serve as laboratory or bench test equipment. Although they share the functionality of some test equipment, it is important that their limitations are understood and taken into account. There are reasons that a quality signal generator can cost thousands of dollars, while an antenna analyzer that includes a signal source costs hundreds. It's not reasonable to assume that each type of instrument can perform the same tasks to the same precision.

Antenna Analyzer as an Impedance Meter

It may be obvious, but it's still worth mentioning that any data that an antenna analyzer can provide about antenna characteristics can also be provided while measuring electrical components or circuits. For example, an analyzer that can measure antenna impedance can also measure the input impedance of a network, or even the impedance of a single component. Usually all that is required is an adapter that provides clip lead connections that mate with the coax jack on the analyzer.

The lack of sign of the measured reactance should not be a problem for measuring the impedance of a capacitor or inductor, since you can likely tell from looking at the component if you are dealing with a capacitor or inductor. Just find a frequency at which the value is not near the maximum or minimum measurement values — measuring values at a few frequencies is a good idea — and make sure that the value changes in the right direction as you shift frequency. Inductive reactance should go up with frequency, while capacitive reactance should go down with frequency. If it goes the other way, you are measuring at a frequency above the self resonant frequency — often a problem with large value capacitors.

If the value of an unknown component is desired, it is just necessary to solve the reactance formulas for the capacitance or inductance value and plug in the analyzer's frequency and measured reactance. For example:

$$X_L = 2\pi fL$$

and

$$X_C = \frac{1}{2\pi fC}, \text{ so:}$$

$$L = \frac{X_L}{2\pi f}$$

and

$$C = \frac{1}{2\pi fX_C}$$

Use caution with units since the basic formulas are in hertz, farads and henrys,

Antenna Analyzer as a Signal Source

All antenna analyzers include an RF signal source that covers their operational frequency range. This signal is available at the analyzer's coaxial connector and can be used directly to inject signals into equipment — just ignore the analyzer's readings, except for the frequency.

Note that I used the term *signal source* rather than *signal generator*. This is not merely a matter of semantics, but rather recognition that the typical signal generator provides not only a stable signal but also an *adjustable output level* that can be used as part of the measurement process. The signal output from most analyzers is a *fixed* level. This deficiency can be made up through the use of an attenuator of some sort, external to the analyzer, and an RF voltmeter or power meter that can read the analyzer's output level — depending on the accuracy of calibration required. Other potential concerns and limitations are described below.

Frequency readout resolution — There are a number of methods of frequency selection employed by antenna analyzers.

The least expensive analyzers tend to have a mechanically tuned variable oscillator with an analog pointer and scale readout. The resolution available is based on the dial increments on the scale and on the thickness of the pointer. With the typical variable capacitor tuning in multiple ranges, the resolution is usually reduced as the frequency increases. The available resolution of such systems often is limited to knowing which amateur band is selected, without knowing the specific frequency within a band. To get a better idea of actual frequency, a receiver of known frequency accuracy can be loosely coupled to the analyzer or circuit being tested. If a frequency counter is available, it can be connected to the output in parallel with the load being tested. In either case, the resolution and accuracy are limited by those of the secondary equipment.

For analyzers using a digital frequency display, either from a frequency counter or a synthesizer, the resolution is simply the number of digits displayed, at least if they stay constant. On some analyzers, particularly on the higher frequency ranges, some of the least significant digits change rapidly because of instability. Arguably, digits that are constantly changing cannot be counted as "resolved."

Frequency accuracy — Some people confuse accuracy and resolution. Resolution is how closely you can read or resolve the quantity measured. (For digital displays, resolution is determined by the number of digits that can be displayed.) Accuracy, on the other hand, is a measure of how close to the absolute standard the value shown on the display is. In the case of frequency, the standard is maintained by the National Institute of Standards and Technology (NIST).

For the case of an analog dial instrument, the accuracy likely will be limited by the resolution of the display in some ranges. Particularly in the higher frequency ranges, the instrument may be limited by the location of the silk screened graduations combined with the care of the alignment technician.

For digital readout instruments, it usually comes down to a question of the frequency tolerance of a crystal controlled oscillator. For those analyzers in which a frequency counter is measuring an oscillator frequency, the accuracy of the crystal controlled time base used by the counter determines the accuracy of the frequency readout. For a synthesized oscillator system, the accuracy of the frequency reference crystal will determine the accuracy of the generated frequency. In either case, the accuracy is significantly better than that of the analog system. For any instrument, a check of the frequency accuracy compared to a frequency counter of known accuracy should let you know what you are dealing with.

Frequency stability — It is important that a signal source used for making measurements stay within the bandwidth of interest in the device being measured or adjusted for the duration of the measurement. Note that the frequency stability needed for making antenna measurements is usually not that stringent — a shift of a few kHz while measuring an HF antenna will not be significant most of the time. If the same source is used for receiver alignment, it could easily drift completely out of the passband.

Waveform purity — Waveform distortion in this context generally means harmonics — spurious signals on multiples of the indicated frequency. A quality signal generator generally has a fairly pure output waveform with most of the energy on its design frequency. Interestingly, in many applications, harmonics will be outside of the range of signals being tested, but they can result in errors in measuring the signal power. On a wideband RF voltmeter or power meter, the power associated with the harmonics can be included in the measurement and make it appear that the signal on the desired frequency is stronger than it actually is.

Antenna Analyzer for Frequency Measurement

A small number of antenna analyzers that use a frequency counter as a frequency display also provide a separate input connector that allows access to the frequency counter function. A switch disables the rest of the unit's functions, turning the analyzer into a standalone frequency counter. This can be a big plus for someone who doesn't have a separate laboratory counter at hand.

Because the analyzer does not usually need to have the frequency accuracy or stability of a laboratory instrument to make useful antenna measurements, it may be that the absolute accuracy of the instrument will not be up to laboratory standards. Still, it can be a useful device and it can easily be checked to find out. Similarly, the frequency stability of the time base may not be up to snuff, but that can be measured to determine the accuracy limits.

Analyzers that don't have a separate input for the counter can still be used to check frequencies using heterodyne techniques. If you have a receiver that has less accuracy or precision than your analyzer, you can tune in the unknown frequency on the receiver, turn off the receiver's beat oscillator and couple the analyzer to the receiver. Adjust the analyzer frequency until you hear a beat note between it and the frequency you want to measure. Carefully tune the analyzer for *zero beat*, the condition in which the sound disappears between the beats on each side. Then the frequency indicated on the analyzer is the same as the unknown frequency.

Antenna Analyzer Simulating a Grid Dip Oscillator

In the vacuum tube era, a popular test instrument was the *grid dip oscillator* (GDO) or grid dip meter. This was basically a vacuum tube oscillator with the inductor exposed on the top of the handheld unit. Most units had multiple plug-in inductors to cover different frequencies, An analog meter was connected to measure the grid current of the vacuum tube. In use, the GDO inductor was lightly coupled to a resonant (L-C) circuit under test. As the frequency of the GDO was varied, at some point the circuit under test absorbed some of the energy from the GDO tuning circuit, resulting in a reduction in grid current — or dip. This occurred at the resonant frequency of the circuit under test.

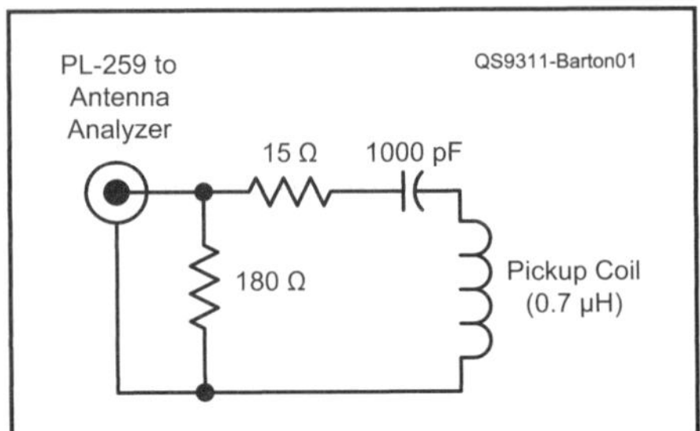

Figure 7.1 — Schematic diagram of the wide range dipper probe for use with an antenna analyzer. The coil is made from three turns of #22 AWG stranded insulated wire, ¾ inches in diameter.

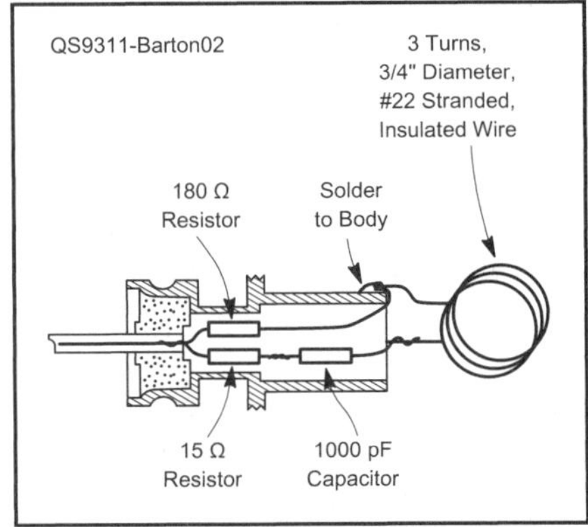

Figure 7.2 — Construction details of the dipper probe. The PL-259 backshell is not shown for clarity, but should be put on before components are soldered to the connector body.

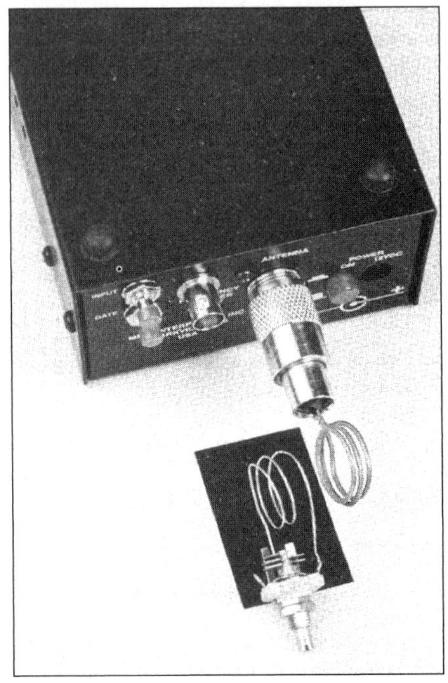

Figure 7.3 — The "grid dip oscillator" probe of Figure 7.2 in use, measuring the resonant frequency of a VHF tuned circuit.

The GDO was a very handy and very popular gadget in its day. As the use of vacuum tubes diminished, some manufacturers offered variations based first on tunnel diodes. Later they made field effect transistor (FET) oscillators. The solid state versions are usually just called "dip meters" but some hams still call them GDOs (perhaps now *gate* dip oscillators). To my knowledge, the only one still in production is offered by MFJ Enterprises, the MFJ-201.

Fortunately, David Barton, AF6S, wrote a *QST* article describing a clever method of using an antenna analyzer as a GDO.[1] While his article was based on a particular analyzer, it should work just as well with any analyzer that measures SWR. His approach was to combine a coupling coil with resistors and a capacitor so that the resulting network would appear to have a moderate SWR over a wide frequency range (see **Figures 7.1** and **7.2**). The operation of the unit takes advantage of the fact that the greater the energy coupled out of the analyzer, the better the SWR looks. Thus if the added link is coupled to a parallel tuned circuit and tuned (see **Figure 7.3**), the SWR indicator will dip at the resonant frequency of the circuit under test.

This arrangement has some decided advantages over a traditional GDO. First, the tuning accuracy of all but the least expensive antenna analyzers is far better than any GDO. Second, the entire frequency range can be covered without having to find and change coils.

DETERMINING TRANSMISSION LINE ELECTRICAL LENGTH

Signals propagating along a transmissions line travel at a velocity somewhat less than the speed of light or radio waves in free space. This is a similar phenomenon to that of light traveling through water — one that we can observe with the naked eye in nature. The reduction in velocity is largely a function of the dielectric constant of the insulating material between the line conductors. Thus air insulated line has a propagation velocity that is just slightly slower than free space, while other insulating materials reduce the velocity more.

This has an effect on the electrical length of the line (that is, the length of transmission line in wavelengths). While we don't care about electrical length for many transmission line applications, it is critical in some cases such as transmission line transformers or phasing and delay lines.

The actual velocity in the transmission line is given by the expression

$$V = V_R \times C$$

where C is the speed of light in free space (approximately 3×10^8 m/s) and V_R is the relative velocity in the line.

V_R is generally provided for each transmission line in vendor data sheets or in handbooks. Alternately, it can be determined if the dielectric material between the conductors is known. In that case,

$$V_R = \frac{1}{\sqrt{\varepsilon_R}}$$

where ε_R is the relative dielectric constant of the insulating material.

As an example, many types of coaxial cable have a dielectric made of solid polyethylene between the conductors. Polyethylene has an ε_R of 2.26, so V_R = 1/1.5 or 0.66. The velocity of signals in the coax will thus be 0.66 $\times 3 \times 10^8$ or 1.98×10^8 m/s, a significant reduction. While vendor data often provides a sufficiently accurate indication of V_R, some cables — especially those with a foam or air-poly mixture — can vary somewhat from lot to lot, or even along a single cable depending on the uniformity of the mix. Thus for critical applications, it is prudent to verify the value.

To determine the electrical length of the line, we could just divide its physical length (measured with a ruler or tape measure) by V_R, but that assumes our cable is made exactly to manufacturer's specification. A better way is to make a measurement of the actual electrical length of the line. We can do this with an antenna analyzer that measures impedance by noting the fact that a line that is half a wavelength (λ/2) long will repeat the terminating impedance at its end, while one ¼ λ long will transform a termination of a short to an open or vice versa.

We make the observation that a reading near 0 Ω on the analyzer is much easier to resolve accurately than one near ∞. Thus, we should plan to measure either a ¼ λ long line that is open at the far end, or a ½ λ long line that is shorted at the far end. The lowest frequency that we measure 0 Ω is the frequency of whichever condition we measure. This is important because the same effect occurs at all multiples of ½ λ and odd multiples of ¼ λ. A physical measurement, or even an eyeball estimate, can usually confirm that you are in the right region.

One source of possible inaccuracy of this measurement is the length of cable or connections between the analyzer's coax connector and the instrument's measurement plane. While this should be minimal, it can easily make a difference for VHF measurements. A correction can be applied, either by checking the apparent electrical length of a known transmission line, or by measuring the length of internal connections, if they are accessible.

CHECKING LINE LOSS FROM ONE END

While measuring loss directly with a source at one end of the line and a power meter at the other is the most accurate method of determining loss, sometimes it is not feasible to get equipment to the far end. In that case we can get a good measure of line loss from one end by measuring the SWR at the frequency of interest with the coax either open or shorted.[2]

If we had ideal lossless coax, our measured SWR would be infinite since all the power would be reflected and would return to the source. With losses in the coax, something else happens. Whatever signal power we put into one end will arrive at the other end with the same power, less the amount of line loss. Because the short or open at the far end of the coax provides a total reflection, the signal then reverses its direction and returns to the source end. On the return trip, the power level at the far end of the coax will be again reduced by the losses in the coax.

The reduced reflected power will show up in real coax as an SWR that is less than infinite. The higher the loss in the coax, the lower the measured SWR as read at the test end. Carried to extremes, a very high loss in a length of coax (such as the losses at UHF for a 500 foot long piece of RG-58) will

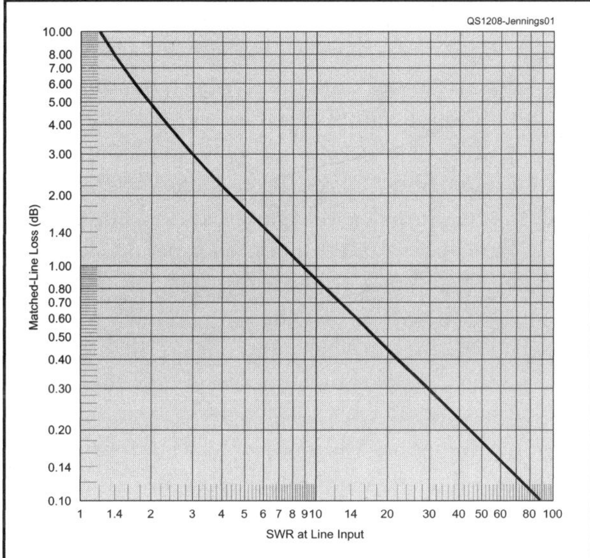

Figure 7.4 — Graph of the apparent SWR versus losses in a coaxial cable terminated in an open or short. Find the SWR value you measured on the horizontal axis and then go straight up and read the one way matched loss in dB on the vertical axis where the SWR reading crosses the diagonal line. Note that if the line is not matched, the loss can increase significantly.

result in an SWR of nearly 1:1 with the far end of the coax open or shorted.

Figure 7.4 shows a graph from the 15th edition of *The ARRL Antenna Book* of the SWR versus matched loss in a coax line measured with a short or open at the far end. The *Antenna Book* also provides an equation to determine the one-way matched loss from the measured SWR: L_M is the one-way matched line loss in decibels at the measurement frequency.

$$L_M = 10 \log_{10} \frac{SWR + 1}{SWR - 1}$$

In other words, if you measure an SWR of 5:1 with a length of open or shorted coax, either Figure 7.4 or the equation indicates that the cable loss will be 1.5 dB at the measured frequency.

To make the measurement, first make sure the antenna end of the coax is not connected to anything, and that any shorts you placed there during ohmmeter continuity tests have been removed. Follow the instructions in the manual for the antenna analyzer to make an SWR measurement at your desired frequency.

As an example, consider the measurement of SWR of a 50 foot length of RG-213 used with a dual-band VHF/UHF antenna. At 147 MHz, the open circuit SWR is 6.6:1. This indicates a loss of about 1.3 dB, compared to 1.4 dB predicted by *TLW* for 50 feet of RG-213 (Belden 8267). At 445 MHz, the measured SWR is 1.4, indicating a loss of 7.78 dB, much worse than the 2.6 dB predicted by *TLW*. Thus this coax, while useful on 2 meters, will not be very good on 70 cm and is probably in the process of degrading.

What if the Antenna is Up in the Air?

The tests described previously are great if you have access to both ends of the coax, but sometimes you would like to find out about the condition of your coax without going up the tower. While you don't have access to the far end, you likely know what type of antenna is there. For example, I have a 2 meter Yagi with a T matching section. While that provides a nice

50 Ω termination at 2 meters, it should look like a short at dc. An ohmmeter continuity test should show a very low resistance, almost as low as if it were a short. A split feed Yagi or dipole should look like an open circuit at dc.

The SWR at 2 meters won't tell us much about loss. If fact, the higher the loss, the better the SWR will look. On the other hand, the 2 meter (144 MHz) antenna will look a lot like a short at 1⁄10 the frequency (14 MHz, or 20 meters). A measurement there should give us a good, but not precise, idea of the 14 MHz loss. If the cable is good there, chances are it is also good at 2 meters. Other types of antenna connections may be trickier to make use of, but if you measure the SWR on other frequencies when the cable is new and store the data in your archives, any change in later years may mean that the antenna or transmission line has undergone a change that merits investigation.

Another trick is to sweep your analyzer frequency looking for the worst SWR. That probably indicates either a very high or very low impedance termination, and the loss at that frequency will be no better than that indicated by the SWR.

A cautionary note — All of the loss results are based on the data provided by the antenna analyzer. Chapter 24 of *The ARRL Antenna Book* notes that the instruments available to most amateurs lose accuracy at SWR values greater than about 5:1, so this method is useful principally as a go/no go check on lines that are fairly long. That really is the only question you need to answer — do we want to keep this coax, or is it time for a replacement? These tests should give you enough information to make those decisions.

DETERMINING TRANSMISSION LINE Z_0

One of the key transmission line parameters is characteristic impedance (Z_0). In most cases, the Z_0 will be identifiable by looking up information on the manufacturer's part number. Occasionally the specified impedance in ohms is written on the cable jacket. Cable sometimes turns up that is unmarked, can't be read or has numbers that don't relate to available information. For most situations, it is necessary to determine the Z_0 so the line's characteristics can be taken into account.

One way to get a good estimate of the Z_0 is to measure the diameter of the inner conductor and the outside of the inner dielectric (the same as the inside of the shield) with calipers. Then, by knowing or estimating the ε_R of the dielectric, the Z_0 can be calculated. Of course, this isn't feasible if the cable has connectors on each end. For coax cable:

$$Z_0 = \frac{138}{\sqrt{\varepsilon_R}} \times \log_{10} \frac{D}{d}$$

where D and d are the diameters of the inner and outer conductors, respectively. For balanced air (or mostly air) dielectric lines

$$Z_0 = 276 \times \log_{10} \frac{2D}{d}$$

In the balanced line case, D is the spacing between the wire centers.

It is also possible to use an antenna analyzer to make the determination. First recognize that the antenna analyzer doesn't really measure SWR, but rather indicates the relationship of the connected load to its design Z_0, usually 50 Ω. As an example, if we had a 75 Ω coaxial transmission line terminated with a 75 Ω resistance there would be no reflection from the termination and the actual 75 Ω SWR would be 1:1. The impedance at any distance along the line would be 75 Ω. Our 50 Ω antenna analyzer, however, can't tell the difference between a matched 75 Ω line and a 50 Ω line with a 1.5:1 SWR, so it will indicate that the Z is 75 Ω and the SWR is 1.5:1.

Note that if we had the opposite case — a 50 Ω line terminated in 75 Ω — the indicated SWR would also be 1.5:1. Arguably, if we had an analyzer that just measured SWR, the two cases would appear identical. The key to determining the difference is watching the impedance measurement as the frequency changes. For the matched 75 Ω line, the impedance will be a constant 75 Ω as the frequency (and thus electrical length) is changed. If we performed the same test with the 75 Ω load terminating the 50 Ω coax, while the SWR would stay constant at 1.5:1, the impedance would change.

At frequencies at which the line is an electrical ½ wavelength or a multiple of ½ waves long, the impedance will repeat the 75 Ω. At frequencies at which the line is an odd multiple (including 1) of ¼ waves long, the impedance will be Z_0/SWR or 50/1.5 = 33.3 Ω resistive.

Thus we can easily find the Z_0 of an unknown coax cable with a single noninductive resistor with a value higher than any expected coax Z_0. Just be sure that the unknown cable isn't so short that you can't tune the measurement frequency to its ¼ wavelength electrical length. That would be about 7 inches if the analyzer goes to 300 MHz.

The procedure is to terminate the coax with a resistor of say 125 Ω, higher than the Z_0 of any typical coax. Change the analyzer frequency and look for impedances with zero reactance. There should be some indicating 125 Ω; these are at multiples of ½ wavelength. Others should be lower, at the ¼ wave frequencies, and these will at Z_0/SWR. For our selected resistor, **Table 7.1** shows the expected value for standard coax that we are likely to encounter. Note that the Z_0 and

Table 7.1
SWR and Minimum Resistive Impedance of Transmission Lines Terminated in 125 Ω

Z_0(Ω)	*SWR*	*R(Ω)*
35	3.6	9.6
50	2.5	20
75	1.7	44
93	1.3	72

SWR in the table are the values associated with the transmission line being measured, *not* the Z_0 of the analyzer.

Another approach is to have loads for each common transmission line Z_0 and try them until you find the one for which the SWR reading stays constant as the analyzer frequency is varied. There is nothing wrong with that approach, other than it requires a box of load resistors rather than just a single load for my method. The matched load method may be most appropriate if you will just need to verify that the Z_0 is a standard 50 Ω and you already have a 50 Ω load available (perhaps intended for other purposes).

CHECKING A FILTER'S CUT-OFF FREQUENCY

In a *QST* article, Wayne Cooper, AG4R, described a way to test the passband and cutoff frequencies of an RF or IF filter designed for use in 50 Ω systems.[3] This technique can be used to check high power transmitter low pass filters, low level receiver band pass filters, or even IF crystal or mechanical filters. **Figure 7.5** shows the general approach.

Measuring the frequencies over which the SWR is close to 1:1 should give a quick verification of the passband frequencies sufficient to know if the filter is working. Unfortunately, other information, such as passband attenuation, skirt steepness or stop band attenuation can't be determined with this setup.

Drawing finer resolution of filter performance requires that the load (in Figure 7.5) be replaced with a load and RF voltmeter or microwattmeter with sufficient resolution to read attenuated levels below the analyzer's output.

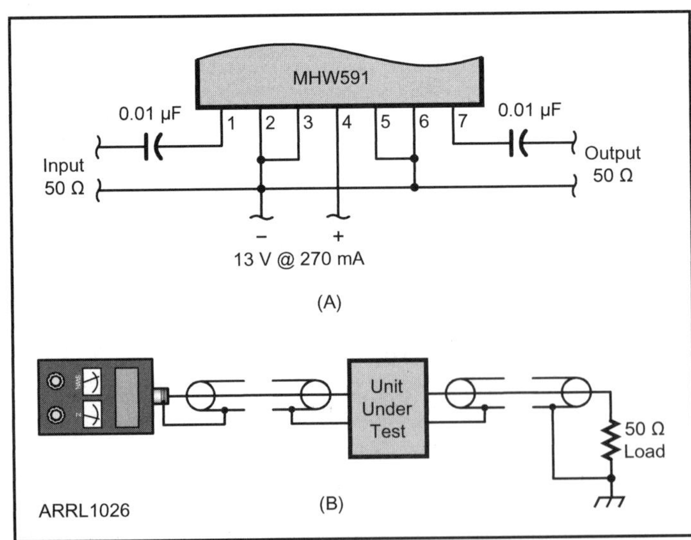

Figure 7.5 — The arrangement shown can indicate the frequencies over which the device under test supports throughput at 50 Ω. If actual passband characteristics need to be determined, the 50 Ω load can be replaced by a 50 Ω RF voltmeter or microwattmeter.

MEASURE HF INDUCTOR Q WITH YOUR ANTENNA ANALYZER

The Q of an inductor is a primary factor in the efficiency of inductively loaded antennas. While Q is defined simply as the reactance divided by the (RF) resistance ($Q = X_L/R$), in real inductors it can be hard to determine because the RF resistance will be quite different from the easy to measure dc resistance. Phil Salas, AD5X, wrote a *QST* article that described a novel adapter he used to measure the RF resistance of components, particularly inductors.[4]

Phil's adapter, shown schematically in **Figure 7.6**, places the inductor under test in series with a variable capacitor. The capacitor is adjusted until the L-C circuit is at resonance at the frequency of interest, as indicated by a resistive impedance reading (0 Ω reactive component) on the analyzer. The increase in the measured resistance value above the 51 Ω reference resistor represents the R value in the Q equation. Or you can use the loss resistance directly to determine antenna efficiency with antenna modeling software.

To complete the determination of Q, just measure the impedance of the inductor by itself at the same frequency. The measured value of reactance equals X_L in the Q equation. Note that at HF, particularly with air dielectric capacitors, the resistive loss in inductors is the major contributor to lowered Q of resonant circuits.

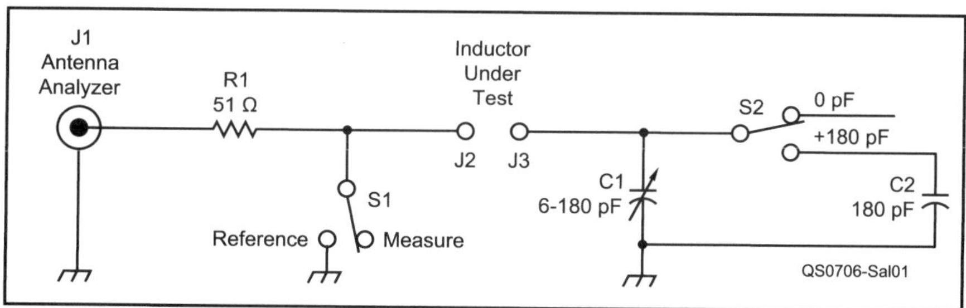

Figure 7.6 — Schematic diagram and parts list of the AD5X adapter for measuring Q with an antenna analyzer. MFJ parts are available from www.mfjenterprises.com, Mouser parts from www.mouser.com.

C1 — 6-180 pF variable capacitor (MFJ 282-5160).
C2 — 180 pF mica capacitor (Mouser 5982-15-500V180).
J1 — SO-239 UHF jack (MFJ-7721).
J2, J3 — Five-way binding post (red, MFJ-606-0003; black, MFJ-606-0004).
R1 — 51 Ω, ½ W film resistor (Mouser 660-CF1/2C-510J).
S1, S2 — DPDT slide switch (MFJ 501-1003).
Aluminum box, 3.25 × 2.13 × 1.63 inches (Mouser 563-CU-3001A).
Double male UHF coax plug (MFJ-7702).
Knob, 1.5 inch diameter (MFJ 760-0125).

A Caveat

It should be noted that at higher frequencies, or if using capacitors with a lossy dielectric (coax line sections, for example), the capacitor losses can also become significant contributors to the total loss. In these cases, the circuit Q is better determined by noting that it equals the resonant frequency divided by the half-power bandwidth and measuring those parameters, perhaps in the test arrangement for filters as described previously. Still, this adapter provides the needed information on losses for most HF loading inductors.

The Details

The adapter is easy to fit into the box called out in the parts list if the specified variable capacitor is used (see **Figure 7.7**). If a different capacitor is selected, check the dimensions before ordering the box for this project. The double male coax connector specified can be used to attach the adapter to the analyzer, or a short coax jumper with UHF plugs can be used.

Wiring is not particularly critical at HF, but it is still important that all connections, particularly chassis ground connections, be solid. The MFJ capacitor specified has front mounting holes that can be tapped for #6-32 machine screws, or #4-40 hardware with lock washers can be used, although they will be tricky to assemble. Other grounds can be connected to solder

Figure 7.7 — Interior view of completed Q meter adapter.

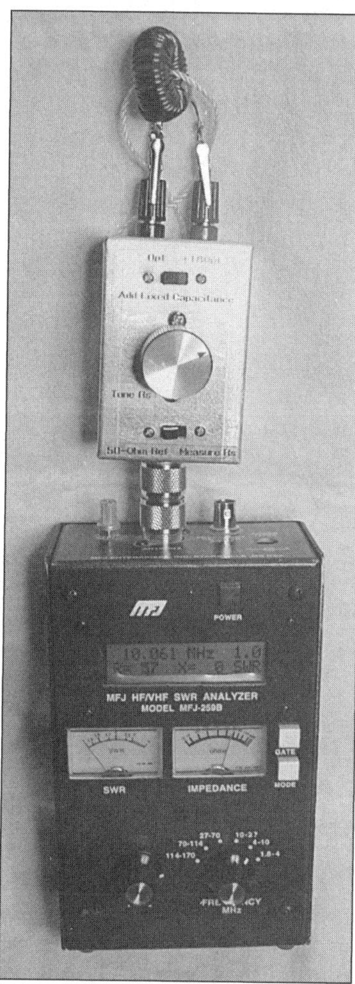

Figure 7.8 — Measuring the equivalent RF series resistance of an inductor.

lugs under the SO-239 #6-32 mounting screws.

Slide switches S1 and S2 require rectangular holes that can be filed from round ones, or a nibbling tool can be used. It would also be possible to use round hole mounted toggle switches, if desired. The labeling shown (see **Figure 7.8**) was made from Casio black-on-clear labeling tape. Alternately, a laser printer and clear film could be used, secured with clear tape.

AN ETHERNET NETWORK TROUBLESHOOTER

An antenna analyzer can even be used as an aid in computer network troubleshooting. Austin McCaskill, Jr, AAØQX, reported in a *QST* article that he was able to use his antenna analyzer to find bad connections in his 10base2 thin Ethernet network wiring without taking down the network.[5] Thin (coax based) Ethernet uses a line with a 50 Ω termination, so a span viewed with an antenna analyzer should indicate a 1:1 SWR. If you are connecting in mid span with a T connector, the parallel connections result in 25 Ω impedance or an indication of a 2:1 SWR. By measuring at a frequency of 10 MHz, Austin found that his test signal didn't interfere with the Ethernet signaling.

In order to find a problem area, he would try wiggling connectors along the LAN structure. If the analyzer reading jumped as he wiggled, he had a suspicious connection.

FINDING THE COAX END WITH THE SHORT

It is always a good idea to make a dc continuity test of a coax cable before putting it into service, since it's much easier to deal with problems before the cable is installed on a tower. The usual approach is to measure the end-to-end resistance of both the center conductor and shield of a cable with fittings. If each shows resistance near 0 Ω, it is making connection. The other test is to measure between the center conductor and the shield — that should show no connectivity or infinite ohms resistance. If a short circuit or low-resistance path is present, the cable is not usable — what to do next?

Joe Wonoski, N1KHB, found himself in exactly that spot with a commercially manufactured cable assembly.[6] He figured that the most likely problem would be a short where the connectors were installed. He could have taken a 50% guess and cut off one connector — if the short went away, he had found it. If not, he had probably cut off a good connector.

Joe correctly reasoned that if he hooked one end of the cable to his analyzer and swept the frequency past the cable's ¼ wave frequency, and if the impedance went very high, that meant the connector at the analyzer end was good and the far end bad. If he did that and it had a low Z at all frequencies, he had the shorted connector end on the analyzer. This is what Joe found and he had to replace the connector at just one end.

Of course, there are other possibilities. If the cable showed a low Z at all frequencies from either end, that meant that both connectors were shorted. If both ends showed a frequency dependent impedance, that would indicate that the short was in mid span. By careful analysis of the lowest frequency that showed a short or a high impedance, one could calculate the likely position of the short along the cable.

Still, the most likely point of failure, in my experience, is at a connector — usually due to melted insulation resulting from the high heat needed to properly solder the shield. Perhaps this is an argument for crimped type connectors.

Notes

[1]D. Barton, AF6S, "An Accurate Dip Meter Using the MFJ-249 SWR Analyzer," *QST*, Nov 1996, pp 45-46.

[2]L. Jennings, WB5IZL, "Quick Coax Test," *QST*, Aug 2012, pp 44-45.

[3]W. Cooper, AG4R, "Antenna Analyzer Tips, Tricks and Techniques: Checking a Filter's Cut-off Frequency," *QST*, Sep 1993, p 38.

[4]P. Salas, AD5X, "Measure Q with your Antenna Analyzer," *QST*, Jun 2007, pp 57-58.

[5]A McCaskill, Jr, AAØQX, "SWR Analyzer Tips, Tricks and Techniques — An Ethernet Analyzer," *QST*, Sep 1996, p 40.

[6]J. Wonoski, N1KHB, "SWR Analyzer Tips, Tricks and Techniques — Locating Shorted Coax Connectors," *QST*, Sep 1996, p 40.

Chapter 8

Enhancing Your Antenna Analyzer

This MFJ-209 antenna analyzer has an adapter inside to provide an audible indication of SWR, in addition to that on the meter. On the outside is an optional audible frequency counter. This arrangement is not only useful for visually impaired users, but also for those who wish to make adjustments without watching a meter or digital display.

As we have discussed, antenna analyzers offer a lot in a small package. They include a combination of test subsystems in a box often small enough to use while hanging off a tower. Still, as with most things in life, there are enhancements that can make them even more useful. Here we will describe a few that others have found helpful.

AVOIDING ACCIDENTLY TURNING YOUR ANTENNA ANALYZER ON AND OFF

There's (almost) nothing worse than climbing to the top of your tower ready to take antenna data and finding your analyzer's batteries are dead or dying. This has historically been a problem with the popular MFJ-259 and similar analyzers because the ON/OFF switch protrudes from the front panel. If the unit is put in a carrying case, or even just sitting about, the switch can be engaged accidentally. The next time you expect to use the analyzer, the batteries are flat.

Fortunately, a simple fix for this common problem was offered by Ed Denton, W1VAK, in *QST* a few years ago.[1] As shown in the upper right of **Figure 8.1**, a rubber grommet is glued to the front panel of the analyzer surrounding the ON/OFF switch. Any size grommet can be used, as long as it meets two criteria:

- The inside diameter must be small enough to be close to the switch, but large enough so that it doesn't interfere with the operation of the switch, when intended.
- The height must be higher than the height of the switch when in the OFF position.

Figure 8.1 — A rubber grommet, higher than the ON/OFF switch on this MFJ antenna analyzer, helps keep it from being accidently turned on.

BALUN FOR MEASUREMENT OF BALANCED SYSTEMS

In Chapter 4, we discussed some of the issues involved in measuring systems using balanced transmission lines to feed balanced loads. For the analyzer to make accurate measurements of such systems, the measurement must be confined to the balanced system itself. Any current flowing from the ground terminal of the analyzer to the cabinet and then via the operator or power supply to ground will be a source of error.

One way to work around the problem is with a common mode choke serving as a *balun* or balanced to unbalanced transition. An effective balun, usable over the HF range, can be made by winding a length of thin 50 Ω coax, such as RG-193, through a ferrite toroidal core such as the FT-240-61 type shown in **Figure 8.2**. The double

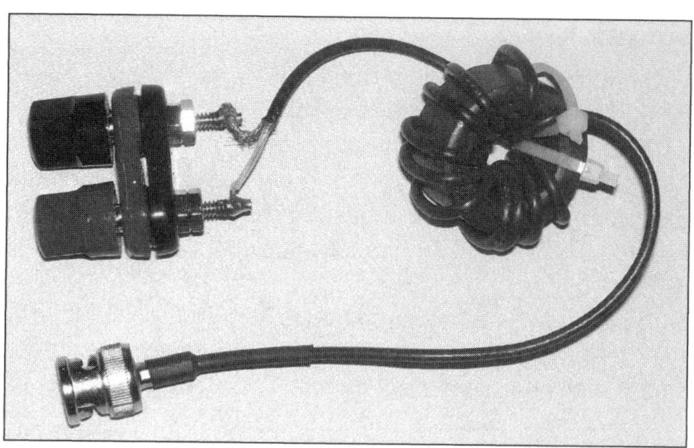

Figure 8.2 — A simple choke balun made by winding multiple turns of coax through a ferrite toroid will have a high impedance to common mode currents. This choke, made of 12 turns of RG-193 miniature coax through an FT-240-61 toroid, is effective from 1.8 to 30 MHz.

banana socket shown uses the standard ¾ inch spacing, a good connector type for balanced transmission lines. If you need a UHF connector, you can use an adapter with the BNC connector shown or obtain a UHF plug designed for RG-193 size coax.

Before winding the balun, measure the total length of coax used. If measuring SWR in a 50 Ω system, the length will not impact the measurement. Note that any other measurement, such as trying to determine the impedance of a mismatched system or of a system at a different impedance (generally the case with balanced transmission line systems), will be changed by a transformation through the differential mode of the short coaxial cable that makes up the choke.

If you are interested in determining something other than the SWR in a 50 Ω system, any of the techniques described in Chapter 6 to transform the impedance from one end of the line to the other can be employed. In this case we are measuring at one end of our short piece of RG-193 and wish to know what the impedance is at the other. This is the reason for wanting to record the physical length of the cable before winding it.

AUDIO READOUT FOR THE MFJ-209 ANTENNA ANALYZER

While the usual visual readout of an antenna analyzer works well for many users, some potential users have difficulty reading visual indicators. This design, adapted from a *QST* article by Tom Fowle, WA6IVG, and Bill Gerrey, WA6NPC, provides a solution using a low cost analyzer.[2] While designed with the vision impaired user in mind, it is also a good tool for those who need to be looking at what they are adjusting, rather than at a meter or digital display.

The only actual modifications to the MFJ-209 are holes in the case ends to accept the machine screws and a grid of holes under the speaker in the back of the case.

Background and General Description

There are two quantities that need to be read from the analyzer — SWR and frequency. Particularly in this example, the SWR is a dynamic value — always changing while adjustments are being made. Making the SWR talk is feasible, but inappropriate, since speech is entirely too slow for the dynamic readout needed while making adjustments. A dynamic audible readout was chosen for SWR readings; this translates the meter reading to a varying audible tone.

In operation, the analyzer's frequency knob is adjusted for the lowest pitch emitted by the VCO (voltage-controlled oscillator) connected to the analog SWR meter. This audible tone is "chopped" (interrupted so as to pulsate) if the SWR exceeds a predetermined threshold. The designers set their threshold at an SWR of 2:1. As the frequency is swept over a band of interest, the tone dips and rises with the SWR. If the SWR is below 2:1, the tone is continuous. When the measurement is above 2:1, the tone is broken or chopped at a rate of about 2 or 3 Hz. With this system, it is very easy to determine the 2:1 band edges and the center resonant frequency of an antenna. Or, a desired frequency can be selected, whereupon the antenna can be adjusted for lowest SWR by listening to the pitch of the VCO.

Determining Frequency

Frequency determination is a bit trickier. The original design performed frequency measurement by connecting the MFJ-209's low-level RF output to the MCount, a Morse output frequency counter that was available from Jackson Harbor Press.[3] See **Figure 8.3**. Unfortunately, the manufacturer no longer has this device available, although they indicated that if there is sufficient demand, they would consider making it available again.

Without audible frequency output, the unit is still worth considering in my opinion. In many applications, the need is to adjust a device for minimum SWR at a particular frequency. For that application, it is only necessary to listen to a receiver tuned to the desired frequency, set the analyzer to that frequency by listening to the receiver and then make the required adjustments for minimum SWR without changing

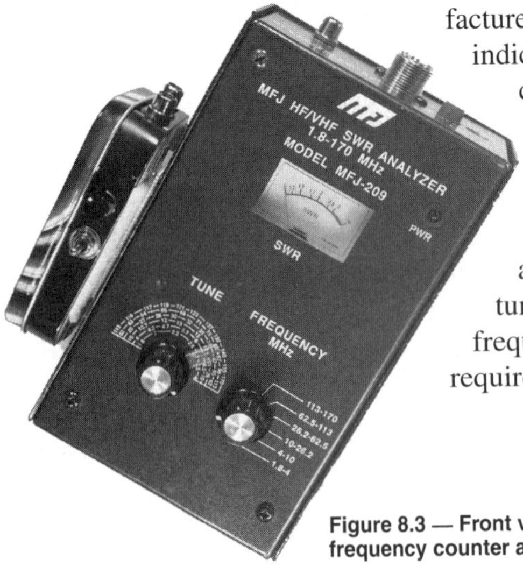

Figure 8.3 — Front view of MFJ-209 analyzer with MCount frequency counter attached at left. [Renee DiVita, photo]

the frequency. Conversely, the frequency of minimum SWR can be determined by tuning the analyzer and then a receiver can be used to find the frequency that the analyzer is tuned to.

If it is necessary to make changes in measurement frequency while tuning (to determine if an antenna is matched over a certain frequency, for example), a receiver can be used to find the analyzer tuning edges. The dial can then be marked at the edges of the desired range with some temporary tactile indicator, such as a strip of light adhesion masking tape.

The Plan of Attack

Some may wonder why the authors of the *QST* article choose to modify the older MFJ-209 analyzer when so many newer digital readout models are available. The answer is that the introduction of digital systems often makes it more difficult to gain access to desired signals to produce non-visual readouts. Although it is sometimes possible to connect directly to digital displays and interpret the information, usually this requires difficult wiring to existing densely packed boards and/or modification of production microcontroller code. The voltage controlled oscillator (VCO) circuit used is popular as a general purpose meter reader.

In the case of the MFJ-209, it is most often only necessary to know the 2:1 SWR points and the center frequency. Therefore the calibration adjustment was left on an internal potentiometer which is trimmed during final installation. In the case of this analyzer, only a few simple connections to the existing unit are required. Using a more expensive digital readout device would have no advantages to the blind technician and the lower purchase cost of the base MFJ-209 may help offset the costs of making these adaptations.

Making it Happen

The original prototype of this adaptation was actually made on an MFJ-249 with a digital counter. This was done simply because that unit was in hand. The two analyzers are nearly identical physically, so this modification can probably be made to many of MFJ's line of analyzers. Our two units were hand wired point-to-point on vector board, since the authors know of no practical techniques by which blind engineers can design PC boards. As with retrofitting any commercial device, the most difficult part of this modification is physically installing the VCO board into the MFJ-209.

The VCO circuit, **Figure 8.4**, including its speaker, is built on a rather Z-shaped piece of vector board that mounts behind the analyzer's main board. The shape of this board, shown in **Figure 8.5**, is such that the batteries

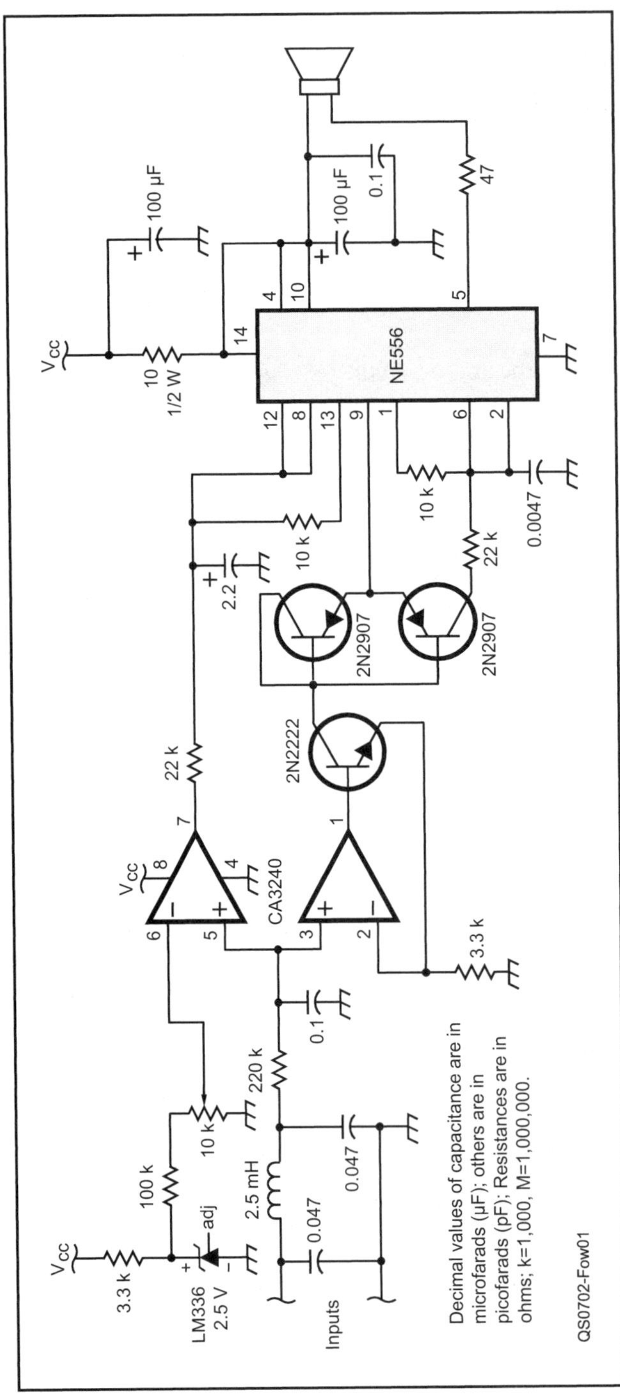

Figure 8.4 — Schematic diagram of the VCO. All parts are readily available standard items.

in the MFJ unit slide past cutouts in our board as shown in **Figure 8.6**, the bottom view of our modified instrument. To support the ends of our added board, there are two custom brackets fabricated from ½ inch aluminum angle stock, tapped for #4-40 machine screws. These were secured to the end panels of the analyzer. If you have a #4-40 tap and the appropriate size drill bit, it is strongly recommended that you tap the mounting holes in the brackets to avoid the difficult task of trying to fit nuts under and inside these brackets.

The vector board is 6½ inches long, just long enough to fit inside the length of the MFJ-209's case. With the case's back and sides removed, and viewing with the connectors up, the upper section of the board is 2⅛ inches square and its right edge is nearly against the right side of the case. The center section of the board is slightly less than an inch wide. The lower section of the board is also 2⅛ inches square and extends to the left of center opposite the upper square section.

The upper section carries the CA3240 op-amp and associated parts while the NE556 IC and its components are strung out down the skinny center. The lower square of the board carries the speaker.

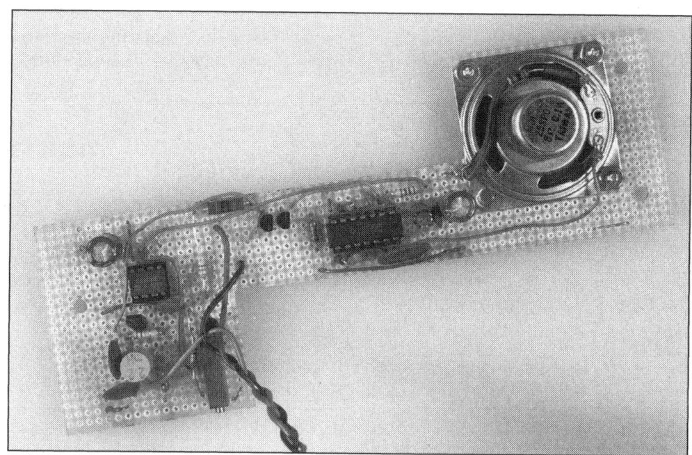

Figure 8.5 — View of the component side of the VCO board before installation in the analyzer. [Renee DiVita, photo]

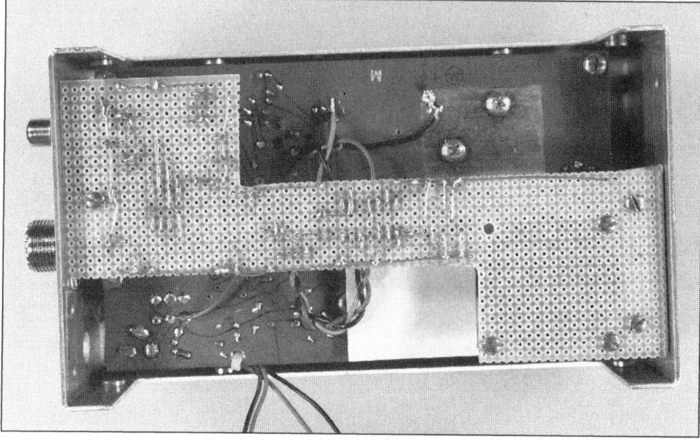

Figure 8.6 — Rear view of MFJ-209 analyzer showing the VCO board installed with its "solder" side visible. [Renee DiVita, photo]

Because it is a close fit, position the board between the battery holders (see Figure 8.6). It is also suggested that the board and brackets be first fitted into the case, making sure that the board passes between the installed batteries before you do the wiring. See **Figure 8.7**.

The upper L bracket is held by one screw that also secures the SO-239 antenna connector. This bracket must be filed a bit to fit around the insulator of the connector. The lower bracket is the width of the board's lower square and carries two #4 machine screws in each side — two holding the bracket to the case and two holding the board to the bracket. Holes to allow the sound from the speaker to escape were drilled in the back panel of the removable piece of the cabinet, positioned adjacent to the loudspeaker. It seems wise to put a business card or other handy insulator between the speaker's magnet and the analyzer's board to avoid shorts. The optional frequency counter, if used, is kept external to the analyzer, connecting through a short phono cable. This leaves the counter available for other uses and reduces the difficulties in modifying the MFJ-209.

Figure 8.7 — Detail of the angle brackets mounted to the bottom of the case and the board interconnections. [Renee DiVita, photo]

Putting it to Work

The optional counter, built in an Altoids box, can be attached to a side of the analyzer case with sticky backed hook and loop material, allowing for a single package but leaving the counter free to go whereever else it is needed. The MFJ-209 offers two other advantages for blind hams:

- The RF oscillator that produces the analyzer's output is modulated at a few hundred Hz. If you don't have the frequency counter available, it is quite easy to find the analyzer's signal in an external receiver, even a portable AM only unit.
- The small pointer on the frequency knob allows quick approximate adjustment as the user gets familiar with band locations on the pointer dial.

The MFJ-209 shown is equipped with a Braille chart showing the ranges of each band. Power for the VCO is taken from the analyzer's 12 V battery and a single wire is soldered to the "hot" meter contact where it protrudes through the main board. This provides input to the VCO. In operation, the VCO comes alive as soon as the analyzer's power button is pressed.

With no load connected to the antenna jack, a high-pitched tone from the VCO will be heard, pulsating as the SWR value goes to maximum. As soon as a load is hooked up, the tone pitch will change to correspond to the SWR. The presence of the chopped tone will immediately indicate a reading of over 2:1. The user simply selects the correct frequency band and sweeps the tuning knob for lowest pitch to find the resonant frequency of the load. A receiver or the MCount is then used to read the actual frequency.

Note that the counter needs to have its prescaler switched out to gain maximum sensitivity below 50 MHz. With a little experience, the blind operator can determine upper and lower 2:1 SWR points and a center resonant frequency within a very few seconds. It is also completely practical to make antenna tuner adjustments and perform all other functions mentioned in the analyzer's manual.

Voltage Controlled Oscillator

The VCO circuit is built around the ancient NE556 timer chip. One half of this timer is the VCO whose charging current is supplied through the first half of a CA3240 op-amp. This op-amp and a 2N2222 transistor make up a voltage-controlled current source that drives a current mirror. This provides the necessary current source referenced to the plus supply line. This configuration results in a VCO with a wide frequency range and very smooth operation. The CA3240 was chosen because its inputs operate properly down to the negative supply rail, which is necessary because the analyzer's meter is referenced to ground. The second half of the CA3240 op-amp is a comparator that compares the incoming meter signal to a calibrated voltage, which is set to reflect a 2:1 SWR reading. The output of this comparator enables the second half of the NE556.

This second oscillator, running at about 3 Hz, turns the VCO on and off, providing the chopping effect when the SWR is high. A temperature-stable LM336 reference diode keeps the calibrated voltage accurate, regardless of battery state. The output of the VCO, the first half of the NE556, drives the speaker through a 47 Ω current limiting resistor. The magnetic speaker can be replaced with a nonresonant type of piezo sounder, eliminating the need for the 47 Ω resistor.

Adjustment

Following VCO board completion and installation in the analyzer, attach a 100 Ω noninductive resistor to the antenna connector. At power up, you should get a tone from the speaker. Turn the calibration pot until you find the point below which the tone is smooth and uninterrupted, and above which the tone begins to pulsate. With the 100 Ω resistor at the antenna connector, the SWR will be 2:1 (100/50 Ω). In other words, a 2:1 SWR indication will now be just on the edge of the start/stop of the chopping.

If you want the chopping of the tone to occur at a different SWR, you can calibrate the instrument by selecting a different test resistor. If a noninductive resistor is connected to the antenna SO-239 connector, the "test SWR" will be the value of that resistor divided by the ideal 50 Ω.

One note of warning about this, and probably most similar antenna analyzers: If you live in a very high RF environment, such as very near a broadcast band transmitter, you may find that on some bands you will never get a good low SWR reading even with an antenna you know to be matched. This is due to the ambient RF being wrongly interpreted by the instrument as "reflected power."

Acknowledgments from the Original Article Authors

The list of folks the authors of the original *QST* article need to thank for help leading to this work is endless but must include: Howard Moscowitz, KB3ZX, for thinking up the modulated VCO years ago. Their dear colleague Albert Alden, *real analog* engineer for the schematic and endless patience and more. Smith-Kettlewell Institute and their boss and friend Dr John Brabyn, for putting up with the authors! Tom's wife Susan Fowle, NY6D, for the original schematic drawing, meter reader board design and everything that matters!

Notes

[1]E. Denton, W1VAK, "Hints and Kinks — MFJ Analyzer On/Off Switch," *QST*, Jan 2007, p 62.

[2]T. Fowle, WA6IVG, and W. Gerrey, WA6NPC, "Audio Readout for the MFJ-209 Antenna Analyzer," *QST*, Feb 2007, pp 35-39. The original article includes a text description of the circuit diagram.

[3]The MCount Morse readout frequency counter kit was sold by Chuck Olson, WB9KZY, of Jackson Harbor Press. Chuck indicates that the chipset is still available, but the kit is not, although he could reissue it if there is enough interest. You can contact Chuck at **wb9kzy@gmail.com**.

Chapter 9

A Survey of Available Antenna Analyzers

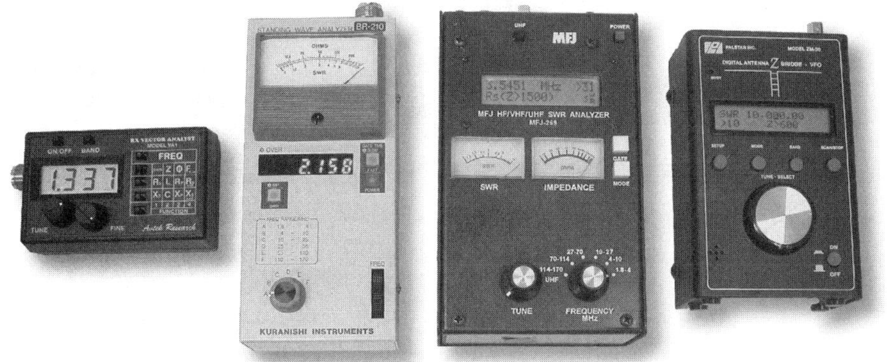

The four antenna analyzers compared in the first group of the chapter are typical in some ways of the variety of antenna analyzers offered.

Talking about antenna analyzers in generic terms is fine, but when it comes down to selecting one, it's important to know what the choices are. In this chapter, we will bring together actual *QST* Product Review information indicating not only the features of different analyzers, but also test data indicating how well the analyzers performed the tasks that they profess to do.

TYPES OF AVAILABLE ANTENNA ANALYZERS

As we have discussed, antenna analyzers come in many forms. While all are capable of measuring SWR, analyzers offer different features both in terms of what they measure and how they display or deliver the results. Some of the key differences to watch for are briefly described here.

What Frequencies Does the Analyzer Cover?

The frequency range of the analyzer should be selected based on the frequency range of antennas or other systems you will want to measure or adjust. All analyzers we have reviewed cover at least the 3 to 30 MHz HF range, the primary operating frequencies of many amateurs. Others extend downward to MF (the 1.8 MHz band or lower frequencies) and some extend upward to VHF and UHF. Frequently, manufacturers offer different models with similar features, but covering different frequency ranges. Not surprisingly, the cost tends to go up as the frequency coverage expands.

Does the Analyzer offer Digital or Analog Displays?

Some analyzers indicate frequency and provide results with analog readouts. There is nothing wrong with this approach, but the resolution is generally much lower than that offered by analyzers with digital displays. Some operators find that it is actually easier to watch the effects of adjustments on analog meters than on digital displays. A few analyzers offer both analog and digital displays — the best of both worlds.

Analyzers that provide for computer output offer a completely different dimension. This can be quite useful for in-shop laboratory type work — bringing a PC to the top of the tower is often a bit of a stretch. Some analyzers finesse the connected PC problem by being able to store data in the analyzer for later analysis. This can be beneficial in many ways.

PRODUCT REVIEW TESTING

The ARRL publishes a Product Review column in its monthly magazine, *QST*. Product Reviews generally include laboratory evaluation of equipment parameters using professional grade, independently calibrated test instruments. Over the years, we have reviewed many antenna analyzers, and I have adapted the results from some of those reviews here. All *QST* Product Reviews are available for download to ARRL members at **www.arrl.org/product-review**. Not all readers of this book will be members, and even ARRL members with Internet access may find it more convenient to compare the analyzers we've looked at recently in one place.

While the reviews in this chapter include equipment that was available at the time of original publication, some models covered are no longer available as new equipment. Even so, it may be of benefit in two ways:

- There is an active used equipment market in Amateur Radio equipment. This may provide a way to knowledgeably purchase an antenna analyzer at a lower cost.
- In some cases, manufacturers have made relatively small changes to equipment, so data on previous models may be directly applicable.

Also, note that prices may have changed.

ARRL Lab Antenna Analyzer Testing Methodology

The ARRL Lab has a standard procedure for the evaluation of antenna analyzers. In addition to taking general data regarding power requirements, output level, frequency coverage and other parameters, the Lab uses each analyzer to measure the impedance of known precision loads. These include values with real (nonreactive) impedances and a sample of complex (resistive plus reactive) impedances. The data tables indicate the sample value as well as the value shown on each of the analyzers.

A LOOK AT SOME HIGH-END ANTENNA ANALYZERS

This review originally appeared in May 2005 *QST*.[1]

Some years back, ARRL Lab Engineer Mike Gruber, W1MG, reviewed the MFJ-207 and MFJ-249 antenna analyzers in *QST*.[2] Versions of those models are still available in MFJ's line, but they have added some new models as well. In addition, three other manufacturers have joined in to offer products in the same general category. In this section, we will review the Autek Research VA1, the Kuranishi Instuments BR-210, the MFJ-269B and the Palstar ZM-30.

The earlier article is available on the ARRL members' website at **www.arrl.org/product-review** and is worth reading if you are not familiar with these handy devices.

A quick summary of the reasons why these instruments are an improvement over just measuring SWR with the SWR meter in your transmitter or antenna tuner might be in order. First, an antenna analyzer allows operation across the spectrum, not just on amateur bands. Second, the power used is miniscule, avoiding unnecessary interference. Third, much more information is available from these units than just SWR. See below to find out the nature of the information; it's different for each model. In addition, they all can also serve as signal generators, and one can even be used as a frequency counter.

Our Test Approach

We tested each unit in the ARRL Lab using some of the same calibrated loads that we use to test antenna tuners. In addition, we added in some samples of complex (reactive) loads. For each load we note the actual value and measured value at a representative set of frequencies. The results are shown in **Table 9.1**. Here are the analyzers in alphabetical order.

Autek Research VA1 Vector RX Antenna Analyst

This is the smallest and lightest of the group (see **Figure 9.1**). It is also one of two units (see the ZM-30 below) that includes the capability to determine if the reactive component of the measured impedance is capacitive or inductive. This unit can make a whole bunch of different measurements selected by button switches in a 4 × 4 array adjacent to the single function display. To use, connect the load to be measured to the UHF connector on top of the unit, push the FREQ button and tune to the desired frequency, then push two buttons to select the desired measurement from the matrix. Once you select a column of the matrix, a single button is all that is needed

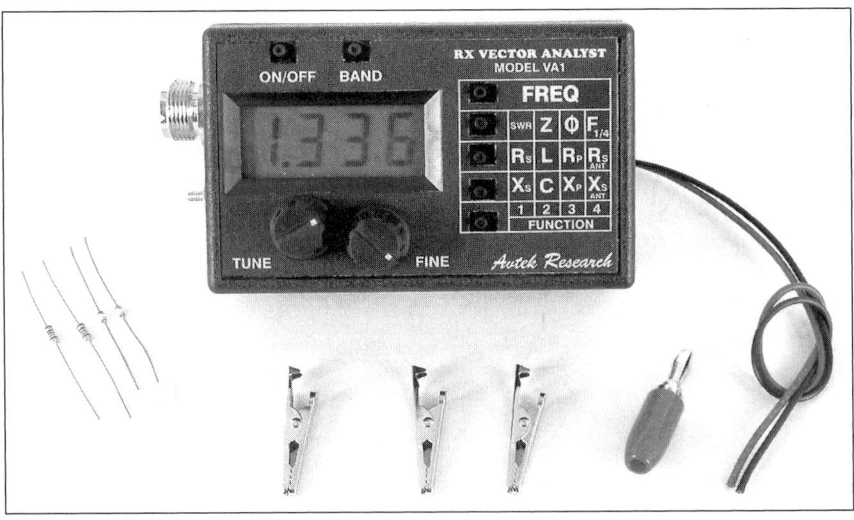

Figure 9.1 — The Autek VA1 antenna analyzer, the smallest and lightest in this group, offers multiple outputs, including the sign of the reactance part of the impedance. This unit allows selection of measured Z_0, not just data for 50 Ω.

Table 9.1
ARRL Laboratory Test Results of Four Antenna Analyzers

Autek VA1 Vector Antenna Analyst

Manufacturer's Specifications	*Measured in the ARRL Lab*
Frequency range: 0.5-32 MHz.	0.44-34 MHz.
Impedance range: 0-1000 Ω.	Wider than 5-1000 Ω.
Impedance accuracy: 20-200 Ω, typically 3-4%.	See Table 9.2.
Warm-up drift: Not specified.	0.42 % in 15 min.
Output power: Not specified.	0.75 mW (50 Ω).
Power requirements: 80 mA (max), 9-12 V dc	70 mA max; measured at 9 V dc.
Size (height, width, depth): 4.1 × 2.6 × 1.5 inches; weight, not specified.	

Kuranishi BR-210 Standing Wave Analyzer, serial number 001478

Manufacturer's Specifications	*Measured in the ARRL Lab*
Frequency range: 1.8-170 MHz.	1.5-172 MHz.
Impedance range: 12.5-300 Ω.	As specified.
Warm-up drift: Not specified.	0.1 % in 15 min.
Output power: Not specified.	0.5 mW (50 Ω).
Power requirements: 160 mA, 8-12 V dc.	320 mA max*, measured at 12 V dc.
Size (height, width, depth): 7.0 × 3.1 × 1.8 inches; weight: 2.0 pounds (with batteries).	

MFJ 269 HF/VHF/UHF SWR Analyzer

Manufacturer's Specifications	*Measured in the ARRL Lab*
Frequency range: 1.8-170, 415-470 MHz.	1.8-175, 415-470 MHz.
Impedance range: Not specified.	>6-400 Ω.
Warm-up drift: Not specified.	0.03 % in 15 min.
Output power: 20 mW (50 Ω).	3.5 mW (50 Ω).
Power requirements: 150 mA (HF/VHF), 250 mA (UHF), 11-18 V dc.	HF/VHF: 160 mA; UHF: 290 mA; measured at 13.8 V dc.
Size (height, width, depth): 6.8 × 4.1 × 2.4 inches; weight, not specified.	

Palstar ZM-30 Digital Antenna Z Bridge

Manufacturer's Specifications	*Measured in the ARRL Lab*
Frequency range: 1-30 MHz	As specified.**
Impedance range: 5-600 Ω.	As specified.
SWR range: 1.0-9.9.	As specified.
Warm-up drift: Not specified.	0 % in 15 min; Freq. accuracy: 6 ppm.
Stability: 50 ppm.	
Output power (50 Ω): 10 mW.	1.0 mW (50 Ω).
Power requirements: 200 mA, 9-16 V dc.	210 mA, measured at 13.8 V dc.
Size (height, width, depth): 5.8 × 3.6 × 2.1 inches; weight, not specified.	

*All button lamps on.
**Tunes 0-30 MHz. Usable down to about 0.1 MHz.

to select other measurement parameters in the same row. This is convenient since the parameters are grouped in a logical way — for example the first column contains buttons for SWR, Rs and Xs (the series equivalent resistance and reactance), the data most likely to be taken. It is also possible to have the measurements cycle between two (or more) parameters. This can be useful if you want to make a series of SWR measurements at multiple frequencies, for example.

In addition to taking basic measurements, the VA1 also performs a number of calculations on the data. Available results are inductance or capacitance values (and you can tell which it is), for complex impedances — magnitude and phase angle of impedance — cable loss determination, based on SWR measurement of open or shorted cable — parallel equivalent resistance and reactance (in addition to the commonly available series values), frequency at which the cable is ¼ λ long — antenna impedance, calculated based on measurement at end of cable. While all these could be calculated off-line with a spreadsheet or calculator, it is handy to be able to determine the derived values right on the display.

Another handy feature is to be able to change the Z_0 of the line being measured from the usual 50 Ω to 25, 50, 52, 54, 73, 75, 93, 95, 112, 150, 300 or 450 Ω. Again, results could be adjusted off line, but having this capability is a real plus, in my opinion.

In use, I found two limitations compared to the other units we looked at. First, the frequency adjustment is rather coarse. The frequency is selected in bands of about a 2:1 range, for example 2.4 to 4.8 MHz or 15 to 32 MHz. The TUNE knob covers the range in just half a turn. A FINE knob is also provided. The FINE knob covers about 10% of the range at the high end and 2% at the low end of a range. Depending on the frequency range, tuning can be quite touchy if you need to take data at a particular frequency. The second limitation is that you can only read one value at a time. It would be nice to be able to watch the frequency change as you look for variation in SWR, for example. The unit will allow you to set it to alternate between the readings, but I found that a bit cumbersome.

The 12 page VA1 manual is quite complete. In addition to clearly describing the operation of the controls and display, it does a good job describing the functions provided and also indicates potential applications in working with antennas and transmission lines.

The VA1 is powered by a single 9 V alkaline battery (not supplied) with a projected life of 6 to 12 hours. There is no direct provision for the use of an external power source, but the manufacturer identifies some aftermarket sources of 9 V battery eliminators that can be used. The unit comes with four precision resistors for calibration use as well clips and wire to make clip leads for connecting to non-coax loads, such as the resistors or balanced line.

Manufacturer: Autek Research, PO Box 7556, Wesley Chapel, FL 33544; tel 813-994-2199, **www.autekresearch.com**.

MFJ-269

MFJ has added to the features of their earlier top-of-the line unit, the MFJ-259, by including a UHF range — from 415 to 470 MHz, covering the 70 cm amateur band with some overlap making it the MFJ-269 (see **Figure 9.2**). The '259 is still available as an MFJ-259B and has all the features except the 70 cm coverage. In addition, MFJ offers a number of lower priced units with reduced, but possibly sufficient, features depending on your requirements.

The '269 does have features! The '269 includes a built in frequency counter, signal generator and impedance measurement system. The '269 has a VHF/UHF oriented Type N connector for load attachment. MFJ also provides a Type N to SO-239 connector adapter for the more common coax connections on HF and VHF. While the generator and counter are clearly designed with the antenna measurement function in mind, they can be used independently for receiver calibration or general measurement purposes. A big plus for this MFJ unit is that all major results are visible at the same time.

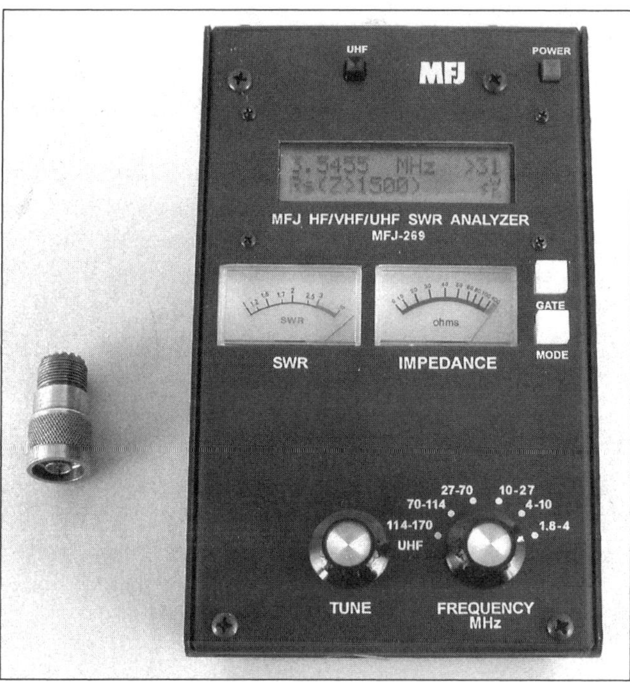

Figure 9.2 — The MFJ-269 antenna analyzer extends the MFJ line into the UHF range and provides considerable flexibility by having external access to its frequency counter.

The two-line LCD display (in impedance measurement mode) simultaneously shows the frequency, R and X values as well as SWR. Analog meters also show the SWR and magnitude of impedance at the same time. While the digital readout provides more precision, the analog meter can be handy when tuning to find frequency of minimum SWR, or to adjust an antenna element length, for example.

The '269 covers the widest frequency range and the tuning is reasonably easy to set. Each band covers about a 2.5:1 frequency range and the knob covers this in three turns. There is no fine tuning, but I was able to adjust to any frequency I wanted without difficulty.

The '269 covers about the same set of derived functions discussed for the Autek unit, some with slight twists. For example, instead of determining the ¼ λ frequency, it can compute the "distance to a short" based on the same information. It also adds a few useful pieces of data, especially *reflection coefficient* and *return loss*. These "advanced modes" provide the requested output while still displaying frequency and SWR on the top line of the display. The analog meters still provide their output in advanced mode as well. Showing six pieces of information at the same time is a real plus.

The '269 will also calculate capacitance or inductance from its reactance measurement. Be careful though. It can't tell which it is, so a given reactance can be converted to either a capacitance or inductance value, but obviously only one is right! If you are measuring a capacitor or inductor, you will know which calculation to make. If you are measuring an unknown load consisting of multiple parts — an antenna and a transmission line, for example — it won't always be easy to tell. The use of a small value capacitive or inductive reactance, compared to the measured reactance, in series with the load should be able to tell you which side you are on.

A well written 38 page manual thoroughly describes the operation of the unit and provides examples of applications.

The '269 operates on 10 internal AA size batteries or an external 12 V dc supply. A wall transformer power supply is offered as an option. By setting an internal switch, the '269 can run from NiCd batteries and recharge them from the external supply. For this function either the optional MFJ supply must be used, or an external supply providing 14 to 19 V is necessary. A 12 V supply will not charge the batteries. An additional function provided is *power saver mode*. If selected, this turns off the display and reduces power consumption by about 90% if you haven't changed anything for three minutes. A poke at the MODE or GATE button will revive it, right where you last left it. A LOW BATTERY indication is also provided.

Manufacturer: MFJ Enterprises, Inc, 300 Industrial Park Rd, Starkville, MS 39759; tel 800-647-1800; **www.mfjenterprises.com**.

Kuranishi Instruments BR-210 Standing Wave Analyzer

This analyzer appears to be a nicely made unit that has a more limited range of measurements available than the other units we looked at. The BR-210 (see **Figure 9.3**) covers a wide frequency range in 2.5:1 bands and has a smooth thumb wheel providing for precisely adjustable tuning with about five turns to cover each range. The frequency is read out on a digital display with a choice of two time bases, one reading to five digits and the other to six. The five digit position is easy to tune to. The six digit display requires very slow

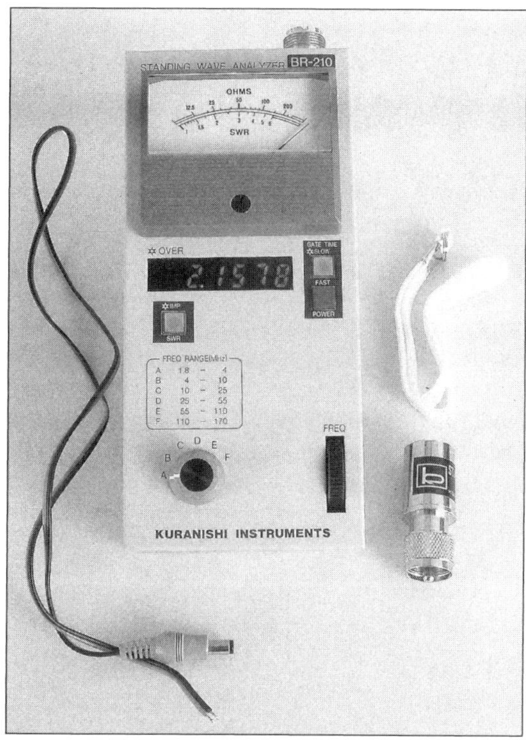

Figure 9.3 — The Kuranishi Instruments BR-210 standing wave analyzer has an easy to use dial arrangement and is good at the tasks it's designed for.

and careful movement to take advantage of the available precision because of the delay between counts of the frequency counter.

The frequency is the only value shown on the LCD display. The other measurements are provided on a 2.5 inch analog meter that can read either SWR or impedance. Unlike the other units, this device only provides the impedance magnitude, not the vector components. The meter shows SWR values to 6:1 with some space above that point, but no additional numbers. The impedance scale is calibrated to 200 Ω in an easy to read geometric scale.

The instructions consist of four sheets of very rough translation into almost English. Fortunately, the operation of this unit is quite straightforward and little coaching should be needed by most users. There are some illustrations of applications included and they are fairly straightforward.

It seems clear that this unit is designed particularly for antenna and matching system adjustment and the taking of SWR data and for those purposes the measurement data is likely to be sufficient for most users. One could also determine capacitor and inductor values by adjusting the frequency to obtain an on-scale impedance reading and calculating off line.

Manufacturer: Kuranishi Instruments (no longer available).

Palstar ZM-30 Digital Antenna Z Bridge

The ZM-30 from Palstar (see **Figure 9.4**) has some unique features. This is an MF/HF only analyzer. It is the only unit that is completely digital, including the VFO, a DDS processor based signal generator. Unlike the other units that have tuning rates that vary with frequency range, this unit can be set to change frequency down to less than 100 Hz per revolution at the smallest step size, or up to 15 MHz per revolution at the highest — a very convenient arrangement.

While measuring impedance, one of its four modes, it shows four data elements on its LCD screen: frequency (to 10 Hz resolution), SWR, real and imaginary parts of the impedance — *with* the sign of the reactance. There is

Table 9.2
Impedance and SWR Measurements of Test Samples Compared to Laboratory Reference

Load	*Frequency*	*Autek RF1*	*Kuranishi BR-210***	*MFJ-269*†	*Palstar ZM-30*	*HP-8753C (reference)**
50 Ω* (1:1 SWR)	3.5 MHz	52–*j*1 Ω (1.0:1)	51 Ω (1.0:1)	48±*j*0 Ω (1.0:1)	53+*j*0 Ω (1.0:1)	
	14 MHz	51–*j*1 Ω (1.0:1)	51 Ω (1.0:1)	48±*j*0 Ω (1.0:1)	52+*j*0 Ω (1.0:1)	
	28 MHz	58–*j*3 Ω (1.1:1)	50 Ω (1.0:1)	48±*j*0 Ω (1.0:1)	53+*j*0 Ω (1.0:1)	
	50 MHz	–	50 Ω (1.0:1)	48±*j*0 Ω (1.0:1)	–	
	144 MHz*	–	50 Ω (1.0:1)	48±*j*1 Ω (1.0:1)	–	
	432 MHz*	–	–	(1.1:1) ‡	–	
5.0 Ω (10:1 SWR)	3.5 MHz	5–*j*1 Ω (9.9:1)	<12.5 Ω (>6:1)	4±*j*2 Ω (12:1)	3+*j*2 Ω (>10:1)	5+*j*0 Ω
	14 MHz	6+*j*0 Ω (8.3:1)	<12.5 Ω (>6:1)	5±*j*0 Ω (9.3:1)	3+*j*2 Ω (>10:1)	5+*j*1 Ω
	28 MHz	5–*j*2 Ω (9.9:1)	<12.5 Ω (>6:1)	4±*j*3 Ω (12:1)	3–*j*4 Ω (>10:1)	5+*j*1 Ω
	50 MHz	–	<12.5 Ω (>6:1)	4±*j*5 Ω (12:1)	–	5+*j*2 Ω
25 Ω (2:1 SWR)	3.5 MHz	25–*j*1 Ω (2.0:1)	26 Ω (1.7:1)	23±*j*5 Ω (2.1:1)	24+*j*0 Ω (2.0:1)	25+*j*0 Ω
	14 MHz	25–*j*0 Ω (2.0:1)	27 Ω (1.8:1)	24±*j*2 Ω (2.0:1)	24+*j*0 Ω (2.0:1)	25+*j*0 Ω
	28 MHz	23+*j*0 Ω (2.2:1)	27 Ω (1.8:1)	23±*j*5 Ω (2.1:1)	25+*j*0 Ω (1.9:1)	25+*j*1 Ω
	50 MHz	–	27 Ω (1.8:1)	24±*j*6 Ω (2.1:1)	–	25+*j*1 Ω
100 Ω (2:1 SWR)	3.5 MHz	100–*j*0 Ω (2.0:1)	100 Ω (2.0:1)	99±*j*17 Ω (2.0:1)	108+*j*0 Ω (2.0:1)	102–*j*1 Ω
	14 MHz	97+*j*5 Ω (1.9:1)	100 Ω (2.0:1)	97±*j*10 Ω (2.0:1)	106+*j*0 Ω (2.0:1)	102–*j*5 Ω
	28 MHz	84+*j*0 Ω (1.7:1)	100 Ω (2.0:1)	95±*j*23 Ω (2.0:1)	102+*j*0 Ω (1.9:1)	101–*j*9 Ω
	50 MHz	–	100 Ω (2.0:1)	87±*j*32 Ω (2.0:1)	–	99–*j*15 Ω

Load	*Frequency*	*Autek RF1*	*Kuranishi BR-210***	*MFJ -269*†	*Palstar ZM-30*	*HP-8753C (reference)**
200 Ω (4:1 SWR)	3.5 MHz	195–*j*16 Ω (3.9:1)	200 Ω (4.0:1)	185±*j*68 Ω (4.1:1)	210+*j*0 Ω (4.0:1)	200–*j*7 Ω
	14 MHz	170–*j*1 Ω (3.4:1)	200 Ω (4.0:1)	183±*j*0 Ω (3.8:1)	205+*j*0 Ω (3.9:1)	195–*j*20 Ω
	28 MHz	147–*j*3 Ω (2.9:1)	190 Ω (4.0:1)	156±*j*86 Ω (4.0:1)	173+*j*56 Ω (3.9:1)	189–*j*37 Ω
	50 MHz	–	190 Ω (4.0:1)	115±*j*98 Ω (3.9:1)	–	175–*j*60 Ω
1000 Ω (20:1 SWR)	3.5 MHz	900–*j*46 Ω (18:1)	>400 Ω (>6:1)	661±*j*743 Ω (27:1)	>600 Ω (>10:1)	978–*j*139 Ω
	14 MHz	590–*j*380 Ω (17:1)	>400 Ω (>6:1)	555±*j*368 Ω (19:1)	>600 Ω (>10:1)	781–*j*405 Ω
	28 MHz	420–*j*11 Ω (8.4:1)	>400 Ω (>6:1)	130±*j*409 Ω (25:1)	104–*j*449 Ω (>10:1)	502–*j*487 Ω
	50 MHz	–	>400 Ω (>6:1)	56±*j*258 Ω (24:1)	–	248–*j*417 Ω
50–*j*50 Ω (2.62:1 SWR)	3.5 MHz	50-*j*47 Ω	80 Ω (2.5:1)	46±*j*47 Ω (2.3:1)	49–*j*47 Ω (2.5:1)	50–*j*46 Ω (2.5:1)
	14 MHz	39-*j*41 Ω (2.5:1)	85 Ω (2.5:1)	63±*j*53 Ω (2.6:1)	44–*j*50 Ω (2.8:1)	48–*j*53 Ω
	28 MHz	55-*j*27 Ω (2.5:1)	80 Ω (1.7:1)	43±*j*45 Ω (2.3:1)	43–*j*43 Ω (2.6:1)	51–*j*48 Ω
50+*j*50 Ω (2.62:1 SWR)	3.5 MHz	54+*j*55 Ω	80 Ω (2.8:1)	50±*j*51 Ω (2.4:1)	55+*j*50 Ω (2.6:1)	52+*j*50 Ω (2.6:1)
	14 MHz	53+*j*54 Ω (2.7:1)	80 Ω (2.3:1)	60 ±*j*42 Ω (2.4:1)	60–*j*51 Ω (2.5:1)	55+*j*49 Ω
	28 MHz	52+*j*34 Ω (1.9:1)	80 Ω (2.3:1)	54±*j*50 Ω (2.6:1)	67+*j*53 Ω (2.5:1)	50+*j*49 Ω

*The SWR loads constructed in the ARRL Lab were measured on an HP-8753C Network Analyzer by ARRL Technical Advisor John Grebenkemper, KI6WX. An HP-11593A precision termination was used for the 50 Ω tests. This termination has a low SWR from dc through the UHF range. The impedance of test loads other than 50 Ω are not accurate above 50 MHz.

**Magnitude of Z indication only, neither R nor X is reported. Readings are approximate as this model only has an analog scale.

†No reactance sign is provided, only magnitude.

‡Only SWR is reported on 432 MHz range.

Figure 9.4 — The Palstar ZM-30 offers a lot of advanced features, including reading the sign of the reactance, over the HF range.

a price to this, however. It determines the sign by shifting the frequency and noting the direction of change in reactance. This takes it a short but noticeable time, so changing frequency needs to be accomplished slowly for the display to keep up. As noted in **Table 9.2**, at one of our sample frequencies, with one of the complex loads, it computed the sign in error. The manufacturer was unable to resolve this, so some caution should be used in taking these results at face value.

The other three modes are inductance, capacitance and VFO. In the first two, you can make the measurements at any frequency you select, and I found quite a variation over frequency with some components in my junk box — good to know. The VFO mode is designed to put out an accurate frequency reference for calibration use, or actual transmitter or receiver frequency control. The fixed output level is specified at ±2 V_{p-p}, or –20 dBm.

Other unique features of this unit include the ability to download software updates from the Palstar website through the serial connector on the unit and the ability to automatically scan across a selected frequency range, looking for a match. A successful search will be indicated on the display, or an audible alarm can be invoked.

I found the ZM-30 quite easy to work with and believe it will be a real contender for those who want a unit for HF only use. The ability to upgrade the software in the field naturally makes me think of features I'd like to see in future releases — one would be to have a choice in impedance mode as to whether the ZM-30 would calculate the sign of the reactance. It would be handy to be able to change frequency more quickly and then go back and spot

Table 9.3
Antenna Analyzer Feature Comparison

Model	*Freq Range*	*SWR*	*Z*	*R*	*X*	*±X**	*Freq digits*	*Counter***	*Calculations*	*Price*
Autek VA1	0.44–34 MHz	15	Y	Y	Y	Y	4	N	Y	$199.95
MFJ-269	1.8-170, 415-470	31	Y	Y	Y	N	5	Y	Y	$349.95
BR-10	1.8-170 MHz	6	Y	N	N	N	5/6	N	N	$429.99
ZR-30	1-30 MHz	10	Y	Y	Y	Y	6/7	N	Y	$399.00

*Ability to determine sign of reactance.
**Frequency counter available for use independent of other analyzer functions.

check for sign as needed.

The ZM-30 is quite well equipped. It is provided with an ac adapter, a double-male BNC adapter and three BNC reference loads as well as a BNC terminated balun for measuring balanced feed line systems. If your antenna systems terminate in a PL-259 rather than a BNC connector, you will need to get an adapter. The descriptive and well written manual runs to 16 pages including calibration and download instructions as well as operation and applications.

Manufacturer: Palstar Inc, 9676 N Looney Rd, PO Box 1136, Piqua, OH 45356; tel 937-773-6255; fax 937-773-8003; **www.palstar.com**.

In Summary

Any of these analyzers could be a worthwhile addition to an Amateur Radio station, or RF laboratory for that matter. They all provide useful functionality but all have different features and specifications, so you'll want to carefully compare your requirements to the specifications and features. Some of the key parameters are described in **Table 9.3**.

Notes

[1]J. Hallas, W1ZR, "A Look at Some High-End Antenna Analyzers," Product Review, *QST*, May 2005, pp 65-69.

[2]M. Gruber, "MFJ-249 and MFJ-207 SWR Analyzers," Product Review, *QST*, Nov 1993, pp 75-77.

ANTENNA ANALYZERS WITH A DIFFERENT VIEW

This review originally appeared in November 2006 *QST*.[1]

The antenna analyzer models discussed in the last section all had advanced features for antenna and transmission line measurements. Some could also serve as signal generators or frequency counters. All were oriented toward displaying a single measurement point at a time.

This section examines two analyzers that provide data output in the form of a graphical display covering a range of frequencies. They can be used alone or with computer software for storing data and for more advanced plotting and analysis. The additional functionality comes with a higher price tag, though.

Why Might You Want One?

These analyzers can — with limitations — provide results at a single frequency, so in that respect they are similar to the devices we've reviewed previously. If you're like me, however, you rarely want data at a single frequency point. Rather, you end up tabulating the data across a range of frequencies or across an entire band. Typically you adjust an antenna or tuner at some mid-range frequency and then confirm it meets requirements over a wider range. Either of these units can show performance over the whole frequency range of interest at one time.

One antenna analyzer application is adjusting an antenna tuner to a frequency before transmitting. Your friends and neighbors will appreciate it if you do. It is also less stressful for both transmitter and tuner if you can perform the tuning adjustments at a very low level from an antenna analyzer before applying full transmitter power. It is very handy to be able to adjust the tuner while looking at the whole band or band segment of interest, rather than just one frequency.

Either of these analyzers can deliver that data, but their sweep speeds are a bit too slow to adjust the tuner controls while observing the results on the analyzer screen sweeping across a range. You need to make an adjustment, then wait for the sweep, then make another adjustment. For most other measurements, such as antenna or stub trimming, you won't notice the sweep time at all.

Compared to an analyzer with a single frequency display, a graphical output saves a number of manual steps and avoids the need to try to read those notes taken while hanging from the tower. Both units have companion software that allows data review and printing using your computer after the fact. An SWR plot of an antenna stored on your computer (with date taken noted) can be an excellent way to confirm proper antenna operation over time.

Our Test Approach

We tested each unit in the ARRL Lab using some of the same calibrated loads that we use to test antenna tuners. In addition, we tested samples of complex (reactive) loads. For each load we note the actual value and measured value at a representative set of frequencies. The results are shown in **Tables 9.4** through **9.6**. Note that, as with other analyzers we've tested, and as discussed below, the sign of the reactance is not always easy to determine. If you need the sign of the reactance, it is our recommendation to verify the sign through alternate means, such as inserting a small (not large enough to change the sign) reactance in series and see which direction the indication moves.

Table 9.4
AEA VIA Analyzer, serial number 0603

Manufacturer's Specifications	*Measured in the ARRL Lab*
Frequency range: 0.1-54 MHz.	As specified.
Frequency accuracy: Not specified.	3500 ppm.
Impedance range: 0-1000 Ω (up to 20:1 SWR).	Better than 5-1000 Ω.
Impedance accuracy: Not specified.	See Table 9.6.
Warm-up drift: Not specified.	7.8 ppm in 15 min.
Output power: 4 mW into 50 Ω.	3.8 mW into 50 Ω.
Power requirements: 150 mA (max), 12-16 V dc	150 mA max; measured at 13.8 V dc.
Size (height, width, depth): 2.3 × 4.3 × 8.5 inches; weight, 1.6 pounds.	

Table 9.5
Timewave TZ-900 AntennaSmith, serial number 10079

Manufacturer's Specifications	*Measured in the ARRL Lab*
Frequency range: 0.2-55 MHz.	As specified.
Frequency accuracy: 1 ppm.	As specified (after calibration).
Impedance range: Not specified.	Better than 5-1000 Ω.
Impedance accuracy: 8%.	See Table 9.6.
Warm-up drift: Not specified.	4.2 ppm in 15 minutes.
Output power: 20 mW, load not specified.	4.3 mW into 50 Ω.
Power requirements: 500 mA, 9-16 V dc.	460 mA while charging battery, 40 mA otherwise.

Table 9.6
Impedance and SWR Measurements

Load	*Frequency*	*AEA VIA*	*Timewave TZ-900*	*Agilent 4291B* (reference)*
50 Ω (1:1 SWR)	3.5 MHz	50+*j* 0 Ω, (1.0:1)	48+*j* 0 Ω, (1.0:1)	50+*j* 0 Ω**
	14 MHz	50+*j* 0 Ω, (1.0:1)	48+*j* 0 Ω, (1.0:1)	50+*j* 0 Ω
	28 MHz	50+*j* 0 Ω, (1.0:1)	48+*j* 0 Ω, (1.0:1)	50+*j* 0 Ω
	50 MHz	50+*j* 0 Ω, (1.0:1)	48+*j* 0 Ω, (1.0:1)	50+*j* 0 Ω
5.0 Ω (10:1 SWR)	3.5 MHz	3.2–*j* 0 Ω, (10:1)	2.4+*j* 0 Ω, (20:1)	5.1+*j* 0.1 Ω
	14 MHz	4.1–*j* 0 Ω, (8.6:1)	2.5+*j* 0 Ω, (20:1)	5.1+*j* 0.5 Ω
	28 MHz	5.5+*j* 2.1 Ω, (9.0:1)	2.3+*j* 2.6 Ω, (22:1)	5.1+*j* 1 Ω
	50 MHz	5.6+*j* 7.9 Ω, (9.1:1)	2.6+*j* 3.0 Ω, (19:1)	5.2+*j* 1.7 Ω
25 Ω (2:1 SWR)	3.5 MHz	25+*j* 0 Ω, (2.0:1)	23+*j* 0 Ω, (1.9:1)	25+*j* 0 Ω
	14 MHz	25+*j* 0 Ω, (2.0:1)	24+*j* 0 Ω, (1.8:1)	25+*j* 0.2 Ω
	28 MHz	25+*j* 0 Ω, (2.0:1)	24+*j* 0 Ω, (1.7:1)	25+*j* 0.5 Ω
	50 MHz	25–*j* 6.7 Ω, (2.1:1)	24+*j* 0 Ω, (2.1:1)	25+*j* 0.8 Ω
100 Ω (2:1 SWR)	3.5 MHz	102+*j* 0 Ω, (2.0:1)	100+*j* 0 Ω, (2.0:1)	102–*j* 1 Ω
	14 MHz	101+*j* 0 Ω, (2.0:1)	101+*j* 0 Ω, (1.9:1)	102–*j* 4.5 Ω
	28 MHz	99+*j* 0 Ω, (2.0:1)	99+*j* 0 Ω, (2.0:1)	101–*j* 8.9 Ω
	50 MHz	94–*j* 11 Ω, (1.9:1)	97+*j* 0 Ω, (1.9:1)	99–*j* 15 Ω
200 Ω (4:1 SWR)	3.5 MHz	199–*j* 0 Ω, (4.0:1)	200+*j* 0 Ω, (3.9:1)	199–*j* 7.7 Ω
	14 MHz	193–*j* 0 Ω, (3.8:1)	193–*j* 26 Ω, (3.6:1)	194–*j* 20 Ω
	28 MHz	176–*j* 44 Ω, (3.8:1)	175–*j* 63 Ω, (3.6:1)	188–*j* 36 Ω
	50 MHz	141–*j* 69 Ω, (3.6:1)	130–*j* 114 Ω, (3.9:1)	174–*j* 58 Ω
1000 Ω (20:1 SWR)	3.5 MHz	940+*j* 0 Ω, (18:1)	979+*j* 0 Ω, (17:1)	983–*j* 119 Ω
	14 MHz	419–*j* 510 Ω, (21:1)	813–*j* 506 Ω, (23:1)	809–*j* 383 Ω
	28 MHz	259–*j* 429 Ω, (14:1)	607–*j* 534 Ω, (22:1)	526–*j* 488 Ω
	50 MHz	131–*j* 238 Ω, (12:1)	171–*j* 633 Ω, (48:1)	267–*j* 430 Ω
50–*j* 50 Ω (2.62:1 SWR)	3.5 MHz	51–*j* 44 Ω, (2.3:1)	60–*j* 46 Ω, (2.3:1)	50–*j* 47 Ω
	14 MHz	47–*j* 47 Ω, (2.6:1)	71–*j* 48 Ω, (2.4:1)	48–*j* 52 Ω
	28 MHz	50–*j* 63 Ω, (2.1:1)	55–*j* 44 Ω, (2.3:1)	51–*j* 48 Ω
50+*j* 50 Ω (2.62:1 SWR)	3.5 MHz	55+*j* 50 Ω, (2.5:1)	65+*j* 51 Ω, (2.5:1)	52+*j* 50 Ω
	14 MHz	61+*j* 50 Ω, (2.4:1)	59+*j* 54 Ω, (2.6:1)	53+*j* 48 Ω
	28 MHz	58+*j* 55 Ω, (2.7:1)	70+*j* 51 Ω, (2.5:1)	65+*j* 51 Ω

*The SWR Loads constructed in the ARRL Lab were measured on an Agilent 4291B impedance analyzer by ARRL Technical Advisor John Grebenkemper, KI6WX.
**An HP-11593A precision termination was used for the 50 Ω tests. This termination is accurate over a wide frequency range.

AEA VIA Analyzer

The AEA VIA Analyzer (**Figure 9.5**) is more than just an antenna analyzer. It can serve as a signal or sweep generator or a field strength meter as well. The manual provides application ideas for measuring the values of inductors and capacitors, trimming transmission lines, determining transmission line characteristic impedance and even acting as a substitute for a "grid-dip" oscillator.

AEA has been at this game for a while. In 1994, *QST* reviewed the AEA SWR-121 HF Antenna Analyst, the first antenna analyzer we had seen with a graphical display.[2] The VIA is a more modern and sophisticated device that covers the frequency range of 100 kHz to 54 MHz. It is part of a family of hand-held test instruments extending to the top-of-the-line AEA Echo vector impedance meter and spectrum analyzer. That model looks similar to the VIA but operates up to 2500 MHz and costs more than six times as much.

Figure 9.5 — The AEA VIA antenna analyzer offers swept frequency analysis.

Packaged in a large hand sized case, the VIA has a monochrome LCD measuring 2.75 × 1.5 inches (see **Figure 9.6**). The controls are all on the front panel beneath the display and include function keys F1 through F5, up down keys for FREQ (frequency) and WIDTH, toggles for ON/OFF and EXAM/PLOT (now used to record to memory), a numerical keypad with digits 0 to 9 and an ENTER key. The controls are all of a good size, even for my fat fingers, and are evenly spaced and labeled so they are easy to read and use. An N connector is used for the antenna or load.

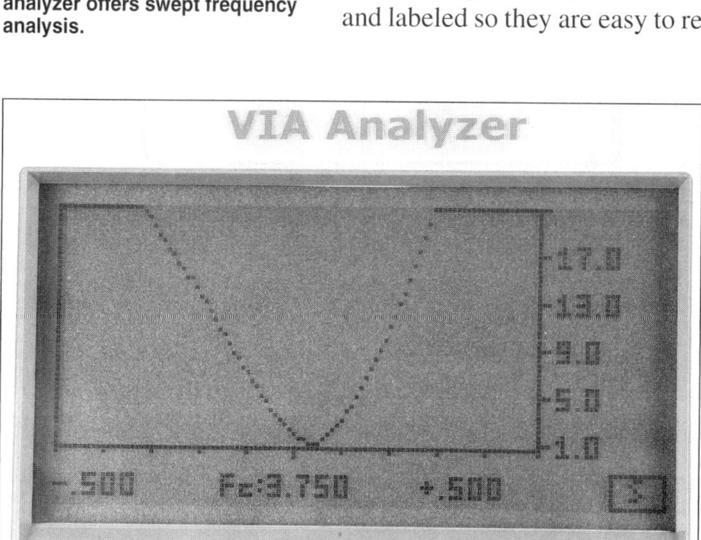

Figure 9.6 — The AEA VIA Analyzer's monochrome LCD screen can display SWR and other parameters over a range of frequencies.

The VIA as a Standalone Device

The VIA can display SWR, return loss, resistance, reactance or a combination of the last two in the form of a Smith chart. Depressing the F5 key cycles through the functions with a small indicator in a box in the screen corner letting you know what you're seeing. The display center frequency can be changed by entering the frequency directly on the keypad or by pushing the FREQ switch up or down. Similarly, the WIDTH control changes the display frequency range.

The VIA is designed to be a handheld field operated device. Data can be read on the screen directly, for field adjustment, or stored for later analysis or archival purposes. My bifocaled eyes found the print on the screen a bit of a challenge to read because of the character size, but it was workable in most light conditions. It operates either from the optional ac supply or from eight AA size batteries.

It is also equipped with a serial port designed to allow it to be used with third-party software. It is in the computer connected mode that the power of the device shines through.

SWR and Impedance Meter

The VIA provides a reasonable display of SWR over frequency as well as the other screens mentioned above. The multiple keys make it easy to shift the center frequency, width or scale of the display among the multiple resolutions allowed. The only functional problem I had was that there was no way to measure single frequency SWR with a rapid enough response update to allow easy antenna matching adjustment. It seemed to take as long to update the display at one frequency as it did looking at the whole band. I was able to make adjustments and obtain good results, but it was not easy.

Sign of the Reactance

As noted previously, these units have limits on their ability to determine the sign (inductive or capacitive) of the reactance. Per communication with folks at AEA, "If the sweep includes a reactive zero crossing, the VIA will not be able to determine the sign. There is zero-crossing distortion, which results in a reactance reading of 0 at phase angles within about 8° of 0°. In addition, for too narrow a plot width, the VIA Analyzer will be unable to resolve the sign but with a plot of greater frequency range (but not near or crossing 0°) it will resolve the sign." They note that their wider range VIA Bravo unit (at three times the price) is much more accurate near 0°.

For many applications, such as trimming antennas or feed lines, this is not an issue. If the data is going to be used to design matching networks, for example, it may be important and these guidelines must be carefully observed and taken into account.

Data Memory

The VIA provides four data memory locations. The key labeled EXAM/PLOT has been redesignated (in our revision 1.2 sample) to serve as a memory storage and retrieval key. Each location contains not just the plot memorized, but all the data taken, so any of the plots from that data sweep can be examined from memory after you get down off the tower.

Sweep Generator

The sweep generator function makes a handy addition to most amateur workshops. It allows a rapid evaluation of filter or amplifier response. It is particularly handy for filter adjustment, since you can observe changes across the whole passband and stop bands while making adjustments. With the VIA in sweep generator mode, the serial jack becomes a synch pulse connection. A 1 µs pulse is sent out the serial port 1.4 ms before each sweep cycle starts. Each frequency is held for 9 ms. At the end of each sweep there is a 16.4 ms interval to allow sweep retrace. The total sweep interval will depend on the sweep width and frequency step size selected. A very nice arrangement.

The output level is fixed at 5 dBm, so a variable attenuator will be needed to set the level for amplifier measurements. A variable attenuator is needed to calibrate the scope's display, to determine where the 3, 6 and 60 dB down points are, so it shouldn't be an issue for someone who needs swept frequency data. The typical setup is shown in **Figure 9.7**.

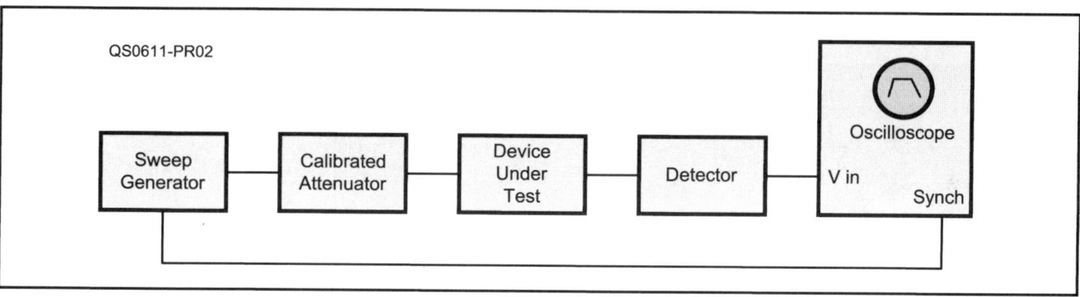

Figure 9.7 — Block diagram of typical sweep generator setup.

Using the VIA as a Computer Sensor

The VIA has a serial port available via a 3-pin jack. Cable wiring instructions are provided, or the supplied 9-pin D-type connector cable can be used. AEA does not offer software directly, but *VIA Director* by Michael Pawlowski, N2MP, is available and is intended to be used with this device. Upon delivery, the software will be in demo mode until you enter the registration code found on the CD case. The computer basically takes over the operation of the unit. You can use the computer to turn the VIA's power on and off and to set any mode or operation.

Figure 9.8 is a representative *VIA Director* screen display, in this case showing SWR over a band. Any of the buttons can be selected to plot additional measured data, or the COMPARISON tab can be used to display any

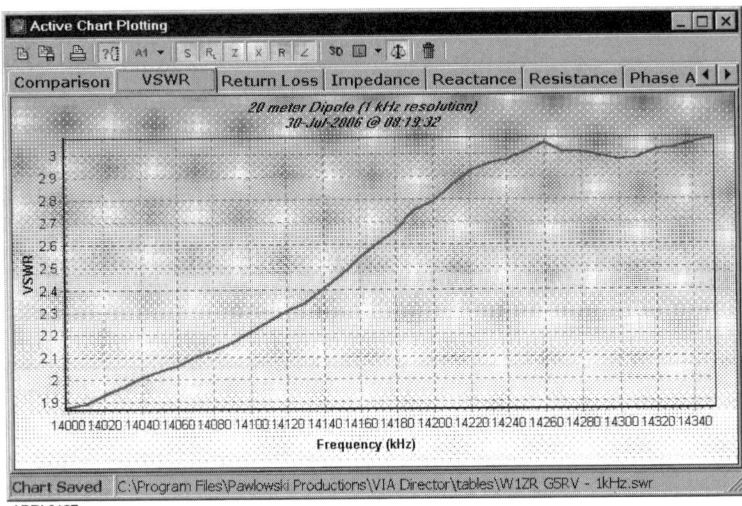

ARRL0127

Figure 9.8 — ***VIA Director*** **SWR plot. Note other choices available in the tabs.**

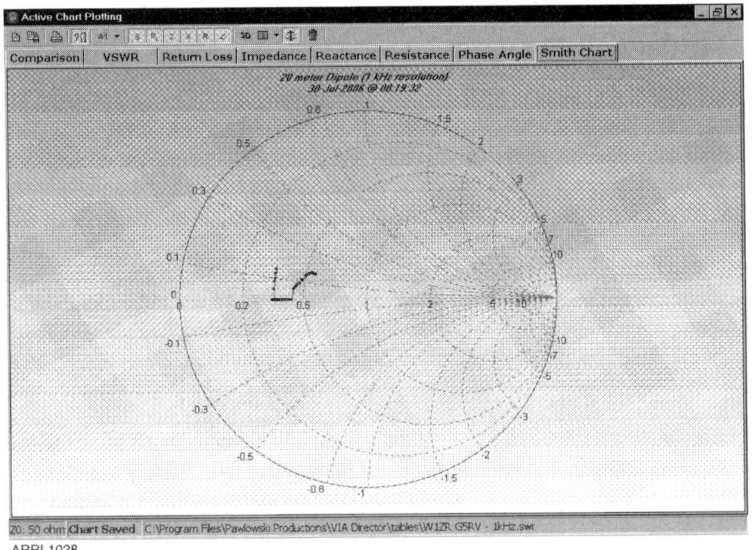

ARRL1028

Figure 9.9 — ***VIA Director*** **Smith chart plot.**

of the different data sets on a common screen. The comparison data is plotted against a single vertical coordinate, so its usefulness can be diminished if there are widely different ranges of values on the different charts.

One choice is the Smith chart representation of complex impedance as shown in **Figure 9.9**. This is a useful way of displaying resistance and reactance on a single graph. The resistance axis is horizontal at the center of the display, with the transmission line characteristic impedance, Z_0 at the center. Lines of constant resistance are circles with one side on the right side of the chart. Reactances radiate from the right center to the edge with inductive reactance on the top and capacitive reactance on the bottom. Thus each complex impedance appears as a single point on the chart. All impedances with the same SWR lie on a circle around the Z_0 point in the center and the impedance found by moving along the transmission line can be determined by progressing around the circle. A change of λ/2 is a full circle around the chart.[3]

Documentation

A very thorough 59 page manual is provided with the unit. In addition to basic operation, it provides details of many applications, including trimming the length of matching sections, measuring capacitor and inductor values and even use as a dip meter to determine the resonant frequency of tuned circuits. The *VIA Director* documentation is in a separate 16 page file included on the software delivery CD.

The unit includes a cable to connect to a computer serial port, an N-to-UHF adapter and a 50 Ω calibration load.

Manufacturer: AEA Technology Inc, 1489 Poinsettia Ave, Suite 134, Vista, CA 92081; tel 800-258-7805; fax 760-798-9689; **www.aeatechnology.com**. Available from some dealers including the K1CRA Radio Store, **www.k1cra.com**. Price: VIA Analyzer, $625.

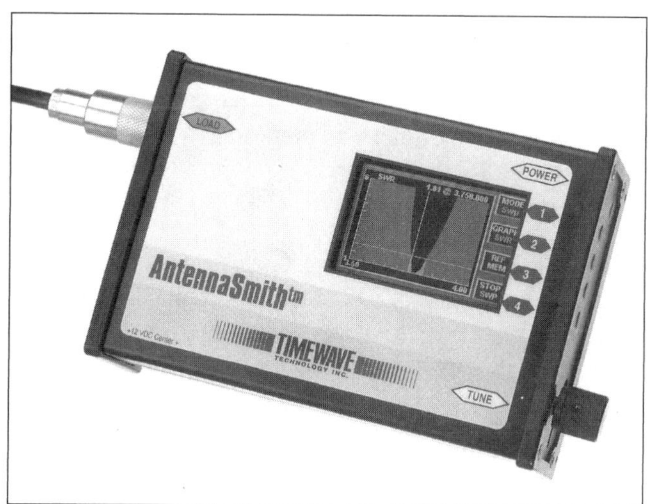

Figure 9.10 — The Timewave TZ-900 is a combination analyzer, data storage unit and direct digital synthesis signal generator, all in a handheld sized unit

Timewave TZ-900 AntennaSmith

The TZ-900 (**Figure 9.10**) is a combination analyzer, data storage unit and direct digital synthesis signal generator in a handheld sized unit that is slightly smaller than the AEA VIA. It operates over the frequency range 0.2 to 55 MHz. Controls include a power switch, frequency control knob and four function buttons that are activated by the unit's control software (see **Figure 9.11**). The output is provided on the front panel color TFT liquid crystal display, including an indication of the functions of the soft buttons. Serial and USB ports are provided for connection to a computer for enhanced display functionality or off-line processing.

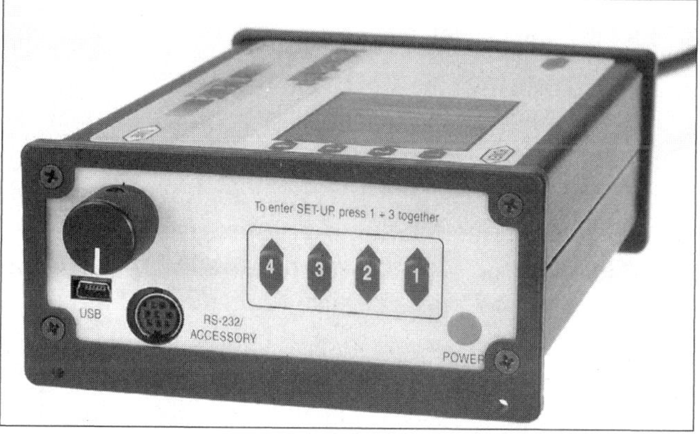

Figure 9.11 — The Timewave *AntennaSmith* controls are mounted on one side of the unit.

The standard antenna connector is an SO-239, but N or BNC connectors are available.

The TZ-900 as a Standalone Device

The TZ-900 can operate as a portable device for field or tower use. While its 2.25 × 1.5 inch display is small by computer standards, it seems more than adequate for its tasks and is readable under most light conditions. The AntennaSmith includes an internal NiMH battery supply that is said to be good for 5 or 6 hours of operation from a full charge. A battery charge indicator appears as a bar graph on the top of the display. The battery can be charged from the supplied ac operated charger or from an external 12 V supply.

The unit provides three major function categories — sweep (SWP), manual (MAN) and synthesizer (SYNTH) modes.

Synthesizer Mode

Starting at the back of the list, the synthesizer function provides a signal from the internal direct digital synthesis (DDS) oscillator right out to the coax connector. The frequency, from 0.2 to 55 MHz, is controlled via the knob on the side of the unit. The display shows the frequency in Hz. On start-up, the knob adjusts the MHz position of the DDS. Push the knob in and it changes to 100 kHz increments and keeps shifting to the right with each push until it reaches the Hz position and then each click shifts 1 Hz. The synthesizer can be calibrated to 10 MHz WWV signals with a single push of a button. The fixed output is specified as 2.8 V_{PP}, or about 13 dBm. It pegged my transceiver's S-meter with a solid, clean sounding signal. This would be useful for receiver frequency calibration, although you may wish to have an attenuator in the line. It also can be used for other signal generator tasks.

Manual Mode

The manual mode acts like many other analyzers — you pick a frequency and it provides the SWR. You can choose rectangular coordinates (real and imaginary parts of impedance) or polar coordinates (phase angle and magnitude). It provides the sign of the reactance or phase angle via an internal algorithm, as best it can determine (see below). The display is provided as a bar graph, a nice idea that may provide for improved visibility and easier tuning while operating in the field. The display can be read in real time, or it can be frozen by the push of a button for later inspection.

I found the response time of the TZ-900, in single frequency mode, to be rapid enough to allow easy adjustment of an antenna tuner. That, combined with its bar graph display, made for an effective tuning aid.

Sweep Mode

The sweep mode allows the display of selected parameters between any two frequencies in the analyzer's range. There are single variable displays (SWR, Z, R or jX vs frequency) or two variable (polar) displays (a Smith chart or a reflection coefficient chart). **Figures 9.12** and **9.13** show two of the presentations possible on the TZ-900's screen.

The sweep ranges can be selected manually or recalled from a selection of 10 memorized ranges. The default values of scan range memories correspond to the 160 through 6 meter amateur bands, just what most of us would want. As with the AEA VIA, in sweep mode the response time is too slow to allow for easy interactive adjustment of an antenna tuner.

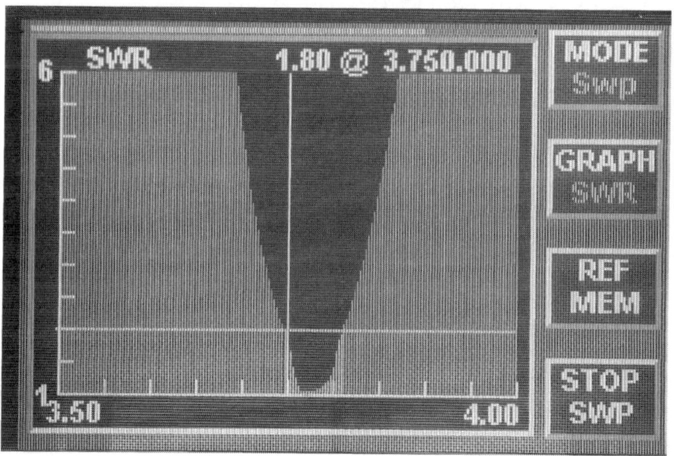

Figure 9.12 — SWR versus frequency for an 80 meter antenna as displayed on the AntennaSmith's color LCD.

In addition to the scan ranges, there are 10 reference memory locations that can be used to store data plots. The stored data can be analyzed off line or compared directly with current data, to note changes made, for example. A name tag can be associated with each reference memory.

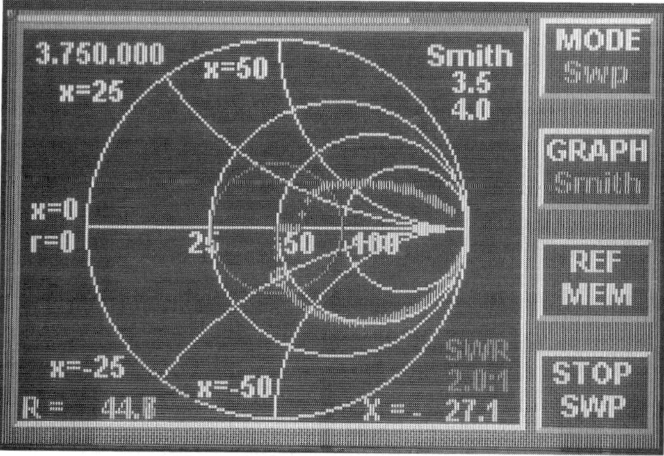

Figure 9.13 — The same load as Figure 9.12 with the *AntennaSmith* switched to its Smith chart display mode.

Sign of the Reactance

As with the AEA VIA, the TZ-900 has limits on its ability to determine the sign (inductive or capacitive) of the reactance. The TZ-900 manual devotes a section to this topic and notes, in summary: "The ...graph(s) depend on the algorithmic determination of the sign of the phase angle. This determination is accurately made in the case of open or shorted transmission lines, antenna resonance measured at the antenna, and most other cases. However, there are instances where multiple resonances occur at closely spaced frequencies or if antennas are measured at the end of random length transmission lines, where the algorithm incorrectly determines the sign of the complex impedance....” The section continues with recommendations for avoiding such problems.

For many applications, such as trimming antennas or feed lines, this is not an issue. If the data is going to be used to design matching networks, for example, it is important and these guidelines must be carefully observed.

Using the TZ-900 as a Computer Sensor

The TZ-900 comes with *Windows* software. Everything it can do as a standalone unit it can do under software control, using the computer as an enhanced storage and display subsystem. Figures **9.14** and **9.15** show two examples. Plots can be recorded to memory for historical purposes or later analysis.

Documentation

Two well written and illustrated manual are included. One, the 37 page *TZ-900 AntennaSmith Hardware Manual,* thoroughly describes the operation of the unit as a standalone device and provides examples of applications. The

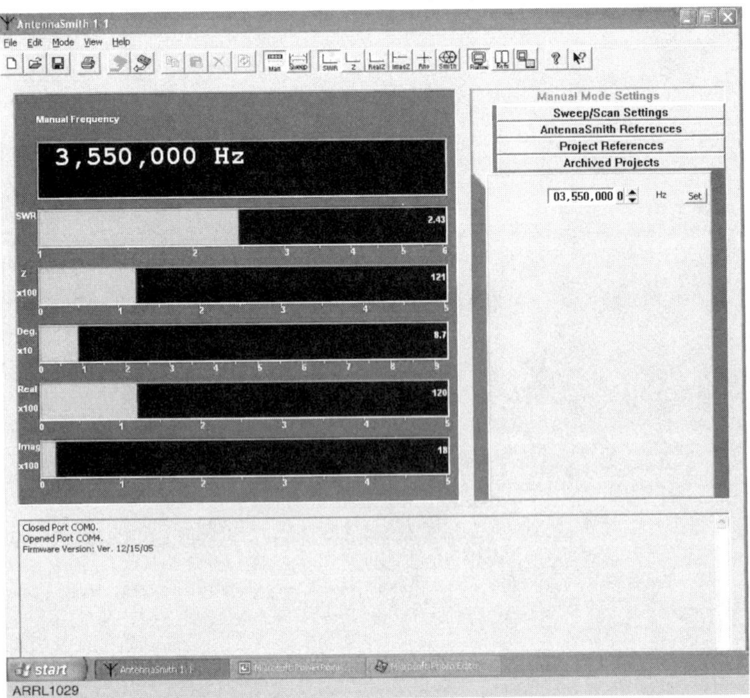

Figure 9.14 — *AntennaSmith* software bar graph tuning chart.

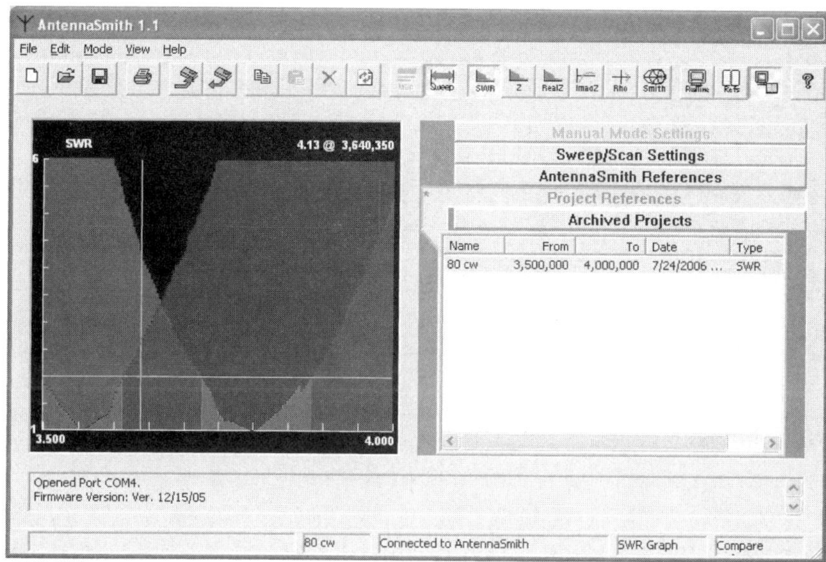

Figure 9.15 — *AntennaSmith* software comparison chart with two stored analysis results.

other, a 54 page *TZ-900* AntennaSmith *Software Manual* — Windows *Version,* provides a comprehensive set of software installation instructions and examples of using the software to store and display data. Both manuals feature color graphics of display and screen presentations that are helpful in visualizing what the system will do. The manual references *Linux* and Macintosh versions, but only the software for *Windows 98SE/XP* appears to be available as we write this.[4]

The unit includes the software, a cable to interface with a computer USB port, an N-to-UHF adapter and a 50 Ω calibration load. A connector for the serial interface is provided, but you need to supply your own cable.

Manufacturer: Timewave Technology, Inc, 1025 Selby Ave, Suite 101, St Paul, MN 55104; tel 651-489-5080; fax 651-489-5066; **www.timewave.com**. *Price*: $900.

Notes

[1]J. Hallas, W1ZR, "Antenna Analyzers with a Different View," Product Review, *QST,* Nov 2006, pp 70-74.

[2]S. Ford, "AEA SWR-121 HF Antenna Analyst," Product Review, *QST*, Nov 1994, pp 77-79.

[3]G. Hall, "Smith-Chart Calculations for the Radio Amateur," Part I, *QST* Jan 1966, pp 22-26 and Part II, *QST* Feb 1966, pp 30-33. Both parts are available on the ARRL Web site at **www.arrl.org/tis/info/chart.html**.

[4]It will not run under plain vanilla *Windows 98*. Fortunately, I was able to borrow a new *Windows XP* laptop from my wife Nancy, W1NCY, for this test. My radio station computer is an elderly *Windows 98* based machine.

ARRAY SOLUTIONS AIM4170: A BENCH TYPE ANTENNA ANALYZER WITH COMPUTER OUTPUT

This review originally appeared in August 2007 *QST*.[1]

The Array Solutions AIM4170 antenna analyzer (see **Figure 9.16**) is an enhanced version of the antenna impedance meter designed by Bob Clunn, W5BIG, and presented in a *QST* construction article.[2] You may want to scan that article before you read further. This model extends the frequency range to 0.1 to 170 MHz from the HF-only version described in *QST*. The unit attaches

Table 9.7
Array Solutions AIM 4170 Antenna Impedance Meter, serial number 202

Manufacturer's Specifications	*Measured in the ARRL Lab*
Frequency range: 0.1-170 MHz.	As specified.
Frequency accuracy: Not specified.	2.3 ppm (after warm up).
Impedance range: 1-2000 Ω.	Better than 5-1000 Ω.
Impedance accuracy: 1 Ω ±5 % 0.1-60 MHz; 10% 60-170 MHz.	See data below.
Drift: 30 ppm.	4.2 ppm in 30 min.
Output power: 20 µW max, load not specified.	15 µW into 50 Ω.
Power requirements: 250 ma, 6-15 V dc.	250 mA at 13.8 V dc.
Size (height, width, depth): 2 × 5 × 4 inches.	
Price: AIM4170, $545; USB adapter and cable, $14.	

Impedance and SWR measurements

Load	*Frequency*	*Array Solutions AIM4170*	*Agilent 4291B (reference)**
50 Ω (1:1 SWR)	3.5 MHz	50.1+*j*0.1 Ω, (1.0:1)	50+*j*0 Ω**
	14 MHz	50.1+*j*0.1 Ω, (1.0:1)	50+*j*0 Ω
	28 MHz	50.0+*j*0.1 Ω, (1.0:1)	50+*j*0 Ω
	50 MHz	50.1+*j*0.1 Ω, (1.0:1)	50+*j*0 Ω
	144 MHz	49.9+*j*0.1 Ω, (1.0:1)	50+*j*0 Ω
5.0 Ω (10:1 SWR)	3.5 MHz	5.0+*j*0 Ω, (9.9:1)	5.1+*j*0.0 Ω
	14 MHz	5.0+*j*0.3 Ω, (10.0:1)	5.1+*j*0.2 Ω
	28 MHz	5.1+*j*0.4 Ω, (9.6:1)	5.1+*j*0.4 Ω
	50 MHz	5.1+*j*0.7 Ω, (9.7:1)	5.2+*j*0.7 Ω
	144 MHz	5.2+*j*1.7 Ω, (9.5:1)	5.2+*j*1.9 Ω
25 Ω (2:1 SWR)	3.5 MHz	25.3–*j*0.1 Ω, (2.0:1)	25.1+*j*0 Ω
	14 MHz	25.3+*j*0.2 Ω, (2.0:1)	25.1+*j*0.2 Ω
	28 MHz	25.3+*j*0.3 Ω, (2.0:1)	25.1+*j*0.4 Ω
	50 MHz	25.2+*j*0.4 Ω, (2.0:1)	25.1+*j*0.7 Ω
	144 MHz	25.5+*j*0.9 Ω, (2.0:1)	25.2+*j*2.0 Ω

to your PC with a serial cable, or to a USB port with an optional adapter and cable. Calibration loads are provided to allow easy setup using menu items on the PC. A BNC coax connector on the front of the unit is used to connect to the sample to be tested, and that's all that happens at the AIM4170, the rest is done from the PC.

What's it Do?

A 61 page manual is included on the CD-ROM provided with the unit, as well as a 10 page *Quick Start* guide. Operation is very straightforward. You

Load	*Frequency*	*Array Solutions AIM4170*	*Agilent 4291B (reference)**
100 Ω (2:1 SWR)	3.5 MHz	101+*j*0.2 Ω, (2.0:1)	100–*j*0.2 Ω
	14 MHz	101–*j*0.5 Ω, (2.0:1)	100–*j*0.9 Ω
	28 MHz	101–*j*0.4 Ω, (2.0:1)	101–*j*1.8 Ω
	50 MHz	101–*j*1.0 Ω, (2.0:1)	99.9–*j*3.1 Ω
	144 MHz	100–*j*3.5 Ω, (2.0:1)	99–*j*8.9 Ω
200 Ω (4:1 SWR)	3.5 MHz	202–*j*0.7 Ω, (4.1:1)	201–*j*1.2 Ω
	14 MHz	202–*j*2.3 Ω, (4.0:1)	201–*j*4.8 Ω
	28 MHz	201–*j*2.1 Ω, (4.0:1)	200–*j*9.4 Ω
	50 MHz	202–*j*4.2 Ω, (4.0:1)	199–*j*16 Ω
	144 MHz	201–*j*12 Ω, (4.0:1)	189–*j*45 Ω
1000 Ω (20:1 SWR)	3.5 MHz	1030–*j*16 Ω, (20:1)	998–*j*33 Ω
	14 MHz	999–*j*19 Ω, (20:1)	981–*j*127 Ω
	28 MHz	986–*j*14 Ω, (20:1)	935–*j*239 Ω
	50 MHz	1000–*j*39 Ω, (20:1)	825–*j*373 Ω
	144 MHz	935–*j*191 Ω, (20:1)	373–*j*476 Ω
50 – *j*50 Ω (2.62:1 SWR)	3.5 MHz	48.5–*j*45.4 Ω, (2.4:1)	50–*j*47 Ω
	14 MHz	47.0–*j*51.2 Ω, (2.8:1)	48–*j*52 Ω
	28 MHz	50.1–*j*47.0 Ω, (2.5:1)	51–*j*48 Ω
50 + *j*50 Ω (2.62:1 SWR)	3.5 MHz	50.5+*j*49.0 Ω, (2.6:1)	52+*j*50 Ω
	14 MHz	51.2+*j*47.5 Ω, (2.5:1)	53+*j*48 Ω
	28 MHz	62.3+*j*50.1 Ω, (2.4:1)	65+*j*51 Ω

*The SWR loads constructed in the ARRL Lab were measured on an Agilent 4291B Impedance Analyzer by ARRL Technical Advisor John Grebenkemper, KI6WX.
**An HP 11593A precision termination was used for the 50 Ω tests. This termination has a wide frequency range.

just tell the software what you want to measure, the desired frequency range and resolution and the type of display — rectangular coordinates or Smith chart — and out it comes.

As noted in **Table 9.7**, the accuracy is exceptional and caused us to go back and check the calibration of our reference loads to make sure we were getting the whole story. Note that while some antenna analyzers are designed to give you a quick answer at a single data point while at the top of your tower, the AIM4170 is really more of a laboratory instrument designed to give you a whole suite of data from the bottom end of your feed line. The manual offers two methods of modifying the output to transform the data to what would be seen at the antenna. There is room for both types of analyzers, in my opinion, in the serious amateur's station inventory.

Figure 9.16 — The Array Solutions AIM4170 is a computer connected bench type analyzer with a number of interesting features.

Figure 9.17 shows the default output, and it looks a bit overwhelming until you decide to select the parameters of interest. Once you deselect the parameters you don't want to look at, it becomes very manageable as shown in **Figure 9.18**. In addition to the plotted data, a click on any frequency will provide you with the tabular data for that frequency on the right hand side of the screen.

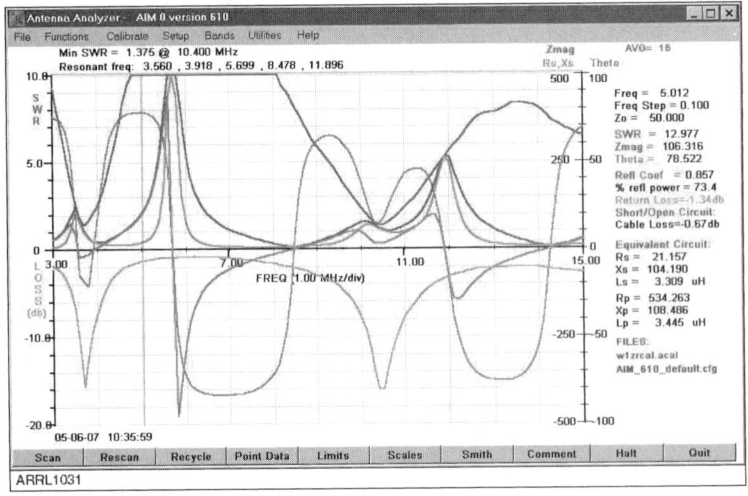

Figure 9.17 — AIM4170 default output with plots of almost all possible outputs in living color.

Setting Your Tuner

A great application for some analyzers is presetting your antenna tuner, avoiding interference on frequency and reducing the strain on your transceiver and tuner. I found that with some practice, and by not

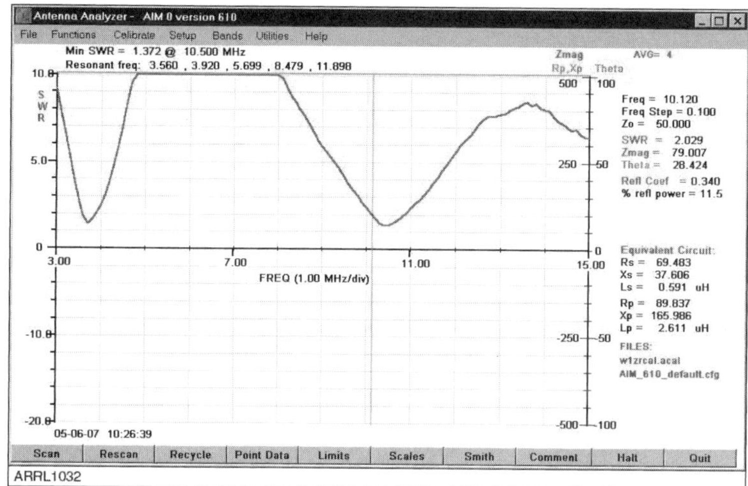

Figure 9.18 — AIM4170 output with just an SWR plot. Seems a bit easier to grasp.

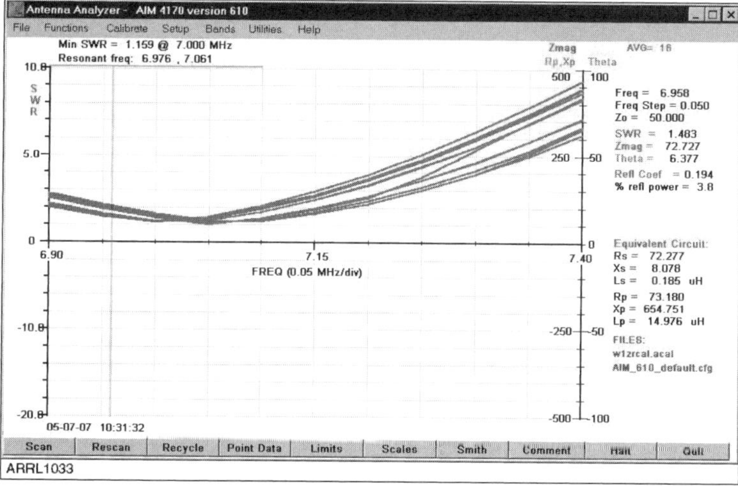

Figure 9.19 — Family of curves as I adjusted my tuner for an SWR of less than 2:1 over the 40 meter CW segment.

having excessive resolution to slow the processing, I could use the sweep function to set my tuner to cover a range of frequencies. **Figure 9.19** shows the family of curves as I adjusted the tuner to have an SWR less than 2:1 over the CW portion of the 40 meter band with my center-fed Zepp.

What Next?

The manual describes many possible applications, including measuring crystal parameters — so don't just think antennas. This is a measurement tool that can be used for a number of other applications as well. For example with a BNC-to-clip lead adapter, it becomes an easy matter to measure the reactance of components to determine their value. Perhaps even more importantly, it is now easy to determine the parallel resonant frequency of an RF choke, or the series resonance of that bypass capacitor, for example.

Did I say crystal? If you're trying to make a crystal lattice filter, select a crystal for your QRP net frequency, or just find out what the story is with that box of strangely marked surplus crystals you couldn't leave behind at a hamfest, just clip one on, select CRYSTAL mode and see the story as shown in **Figure 9.20**. One caution, the analyzer asks for the crystal frequency before it starts. If you put in a frequency very different from the actual frequency,

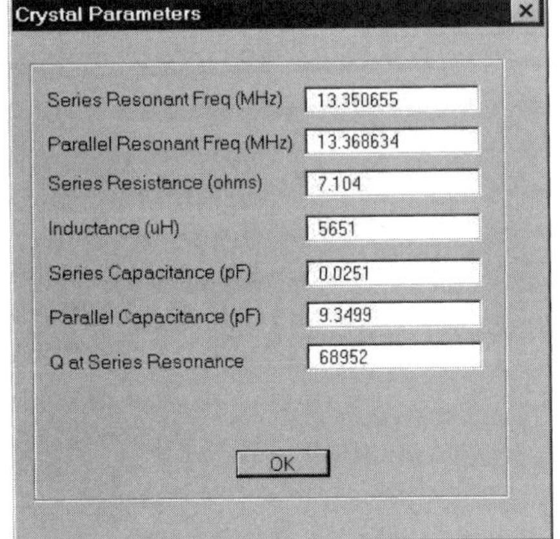

ARRL1034

Figure 9.20 — AIM4170 output while measuring crystal parameters.

it will give you data that looks real but isn't. The easy solution is to do an impedance scan at narrow resolution (I found 0.010 kHz steps worked well) and the real resonant frequency will be obvious.

The AIM4170 software and documentation has undergone several updates and revisions since introduction, and new versions are available by download. It's worth checking the Array Solutions website periodically to see what's new. All in all, this appears to be a very useful and accurate station accessory.

Manufacturer: Array Solutions, 350 Gloria Rd, Sunnyvale, TX 75182; tel 972-203-2008; **www.arraysolutions.com**

Notes

[1]From J. Hallas, W1ZR, "Three Antenna System Measurement Devices," Product Review, *QST*, Aug 2007, pp 65-69.

[2]R. Clunn, W5BIG, "An Antenna Impedance Meter for the High Frequency Bands," *QST*, Nov 2006, pp 28-32.

A LOOK AT FOUR MORE ANTENNA ANALYZERS

This review originally appeared in March 2012 *QST*.[1]

Two of the units we are reviewing, the MFJ and RigExpert are updates to analyzers in existing product lines, while two, the Comet and TEN-TEC are new entries into the field. While similar at first glance, there are significant differences among the units that provide clear choices depending on your interests.

Comet CAA-500 Standing Wave Analyzer

Comet, a longtime antenna and accessory manufacturer, has joined the antenna analyzer marketplace with the Comet CAA-500 (see **Figure 9.21**). This unit measures SWR and magnitude of impedance across the widest frequency range of the units in this test. Our unit covered 1.53 to 508 MHz in 7 overlapping ranges except for a somewhat surprising gap from 259.4 to 273.4 MHz. While the manufacturer indicates that the unit can measure SWR from 1:1 to ∞, there are no numbers above 6:1, so readings above 6:1 are indications, but not quite measurements, in our view.

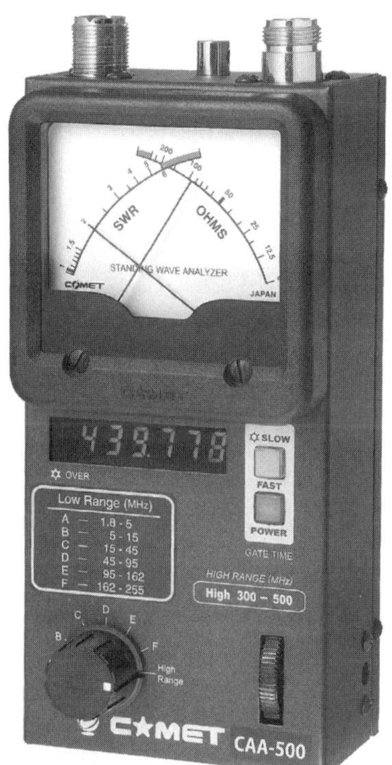

Figure 9.21 — The Comet CAA-500 antenna analyzer measures SWR and magnitude of impedance across the frequency range from 1.53 to 508 MHz, with the exception of a 14 MHz gap around 266 MHz.

The frequency is adj usted using a thumbwheel knob that can be tuned by the same hand that holds the unit — very convenient if you are also hanging onto a tower or ladder, for example. The frequency is indicated on a seven digit frequency counter, while the SWR and impedance magnitude are shown on a large two-scale cross-needle analog meter. There are two coaxial connectors provided, a UHF (SO-239) type for the lower 6 bands and an appropriate Type N socket for the highest range, 225 to 500 MHz.

The unit can be powered by six internal AA size alkaline batteries or an external 8-16 V power supply via a coaxial power connector. A power cable with matching plug and pigtail leads is provided as is a handy wrist strap.

On the Bench

We found that the unit had excellent frequency accuracy, within a few Hz, and easy setability, especially on the lower bands. On the highest band, it was difficult to set it within 50 kHz, although for antenna work in that range that shouldn't frequently be a problem. The frequency stability had similar results, quite steady in the lower ranges, but we observed drift of 250 kHz during 5 minutes at 440 MHz. The frequency counter on the unit follows the drift so you know the actual frequency as it moves.

The frequency counter has two positions, SLOW and FAST, set by a front panel button. In the FAST position, which easily follows tuning, at least on the lower bands, the counter resolution is 1 kHz. In the SLOW position, it reads to 100 Hz, dropping the hundreds of megahertz digit on the higher bands.

The steady output level makes it a natural for use as an accurate signal generator for receiver alignment. The addition of a calibrated step attenuator would result in a test instrument that could be used for sensitivity measurement. If testing the receiver portion of a transceiver, be sure to disable the transmitter to heed their warning about applying RF power to the unit. While we didn't test this analyzer as a dummy load, I can almost guarantee it won't make it.

The impedance meter can also be used to measure the reactance of a capacitor or inductor, as long as you know which it is. Change the frequency until you have the meter in an easy to read region and you will know the reactance at that frequency. By use of the appropriate reactance formula, you will know the value of the capacitor or inductor. Measured performance is shown in **Table 9.8**.

Table 9.8
Comet CAA-500 SWR/Impedance Analyzer

Manufacturer's Specifications	*Measured in the ARRL Lab*
Frequency range: 1.5-500 MHz.	1.532-259.4 and 273.4-508 MHz.
SWR measurable range: 1.0-∞.	As specified, numerical indication to 6:1.
Impedance range: 12.5-300 Ω.	As specified.
Impedance accuracy: Not specified.	See Table 9.13.
Output power: 0 mW (0 dBm) max. load not specified.	0.5 mW (–3 dBm) into 50 Ω at 14 MHz. 0.44 mW (–3.5 dBm) into 50 Ω at 144 MHz. 0.59 mW (–2.3 dBm) into 50 Ω at 440 MHz.
Power requirements: 8-16 V, <180 mA.	165 mA at 13.8 V dc (external power); 167 mA at 9 V dc (internal batteries).
Size (HWD): 7.5 × 3.6 × 2.5 inches, weight 1.75 lb.	

Documentation

The CAA-500 comes with a clearly written four page *Instruction Manual* that includes specifications, identification of each connector and control and a short discussion of how to use it. There are also some frequently asked questions (FAQ) that may be helpful. While not a lot of information is provided, the operation of the analyzer will be intuitive to most amateurs who knew they wanted to buy one.

Bottom Line

This is an easy unit to like. Within the limits noted, it is easy to set and easy to read while making antenna or tuner adjustments. It doesn't offer all the measurement capabilities and other functionality of some of the other units, but it does what it does quite nicely.

MFJ-266 HF/VHF/UHF Antenna Analyzer

MFJ arguably offers the widest selection of antenna analyzers known to man. They have models covering a wide price and capability range starting with their entry level HF analog tuning, analog SWR only metering unit at under $100, ranging up to the MFJ-269PRO HF/VHF/UHF multifunction digital display meter in the $400 range. A look at our Product Review archive (**www.arrl.org/product-review**) will find reviews of a number of representative models.

The MFJ-266 (see **Figure 9.22**) falls near the higher end of their product line and includes many features of the MFJ-269 at a lower price and in an entirely new, more compact envelope with a different control layout. Features include the capability to measure not only SWR, but also the magnitude of

Figure 9.22 — The MFJ-266 antenna analyzer includes many features of the previously discussed MFJ-269 at a lower price and in an entirely, more compact envelope.

impedance and the rectangular resistive and reactive values. The two-line LCD simultaneously shows the frequency, tuning band, complex impedance, magnitude of impedance and SWR — no need to change settings, it's all there. Note that while a plus sign is shown with the reactive component, they describe it as a "place holder." You will need to determine the actual sign by other means such as changing the frequency slightly and noting the direction of reactance change.

Very useful features beyond the comprehensive SWR functionality mentioned above include the use of the '266 as a frequency counter. You enter frequency counter mode by selecting the appropriate buttons of the BAND button set and the DOWN button after power up. In addition to the observed frequency, the display shows the relative strength of the signal. This can be useful to identify a strong received signal that could interfere with antenna measurements.

By pressing the UP button at power on, the '266 will measure capacitance directly in picofarads. Similarly, pressing the DOWN button at power on switches to inductance measurement mode — both very handy features — calculator not required! Again, you need to know which type it is to get the correct answer.

Setting Up the MFJ-266

When the '266 is powered up, the display prompts you, once you know the code, to tell it what you want, starting with the BAND-MODE SELECT buttons. If you press both the UP and DOWN buttons immediately on power up, it will turn on the backlight — the default is BACKLIGHT OFF to conserve battery power. The available dc voltage is shown, along with an indication that you should push UP to select frequency counter mode or DOWN to select antenna analyzer mode.

Frequency is selected from the eight bands by first using the A and B buttons to select HF, VHF, UHF or COUNTER as indicated in the table next to the buttons. While the VHF (85-185 MHz in their definition) and UHF (300-490 MHz) ranges are tuned in one band each, their HF range (1.5-65 MHz) is covered in six bands selected by the UP and DOWN buttons identified as BAND-MODE SELECT in the unit's center. Once you select the range, you tune the frequency using the TUNE knob. The TUNE knob is part of a 10-turn assembly that permits fine adjustment, but it is tricky to set the exact frequency you want, especially on the higher bands. Interestingly, on the "HF" bands, turning the knob clockwise *decreases* the frequency, while on the VHF and UHF bands it works the other way. Note that all US amateur bands from 160 meters to 70 cm are covered, except the 222 MHz band — possibly an

Table 9.9
MFJ-266 Antenna Analyzer

Manufacturer's Specifications	*Measured in the ARRL Lab*
Frequency range: 1.5-71, 85-185, 300-490 MHz.	1.52-71.7, 85-185, 248-530 MHz.
SWR measurable range: 1.0-9.9:1	As specified.
Impedance range: Not specified.	5-200 Ω.
Impedance accuracy: Not specified.	See Table 9.13.
Output power: 1.6 mW (+2 dBm), load not specified.	2.3 mW (+3.6 dBm) into 50 Ω at 14 MHz; 1.6 mW (+2.0 dBm) into 50 Ω at 144 MHz; 0.8 mW (+0.7 dBm) into 50 Ω at 440 MHz.
Power requirements: 10.8–13 V dc (maximum), current, not specified.	Analyzer mode, backlight on, 152 mA; backlight off, 126 mA; field strength mode, backlight off, 41 mA, all at 12 V dc.

Size (HWD): 6.8 × 4 × 3.2 inches, (incl protrusions); weight, 1.3 lb.

important omission for those working that band. Even though the specifications (and band switch) allowed for a gap from 65 to 85 MHz, our unit covered up to 72 MHz, nicely extending through the UK 4 meter band (70-70.5 MHz). Lab measured performance is summarized in **Table 9.9**.

Power Requirements

A somewhat surprising external power requirement is worth noting. The manual states that the external dc supply (plugged into a front panel coaxial jack) needs to be between 10.8 to 12.5 V and offers a warning that it can't be higher than 13 V without load. Since most amateur station dc power supplies put out 13.8 V or more, this may be a problem for some applications, unless special care is taken. The manual also notes that the usual rechargeable 1.2 V NiCd cells will not provide enough voltage for operation. Earlier units could operate from 11 to 18 V.

Documentation

The MFJ-266 comes with a 20 page instruction manual that includes not only instructions but also application notes on how to perform the many tasks that this analyzer can accomplish. The instructions are well written, clear and will be needed to be able to make best use of the unit and all its capabilities.

Bottom Line

The MFJ-266 can serve as the "Swiss Army Knife" in your Amateur Radio tool kit. Either by itself, or with available options it can perform many functions to keep your antennas and station equipment at peak performance.

RigExpert AA-54 Antenna Analyzer

The AA-54 (see **Figure 9.23**) is a very different type of unit than the other three analyzers in this report, although there is a large functional overlap. The AA-54 is part of a family of analyzers that cover different frequency ranges, the upper limit of each identified in the numerical portion of the model designator. Now included in the series are the AA-30, 54, 230, 230PRO, 500, 520 and AA-1000. As you might expect, the price increases, as do the features, as you move up the list. They are described and compared on the RigExpert website (**www.rigexpert.com**).

The first difference you would encounter between using the AA-54 and the other units in this review is that instead of a tuning knob to select frequency, there is a keypad. This is a mixed blessing in a way — it takes a bit longer to fine tune frequencies, but the frequency you get is the one you actually want and it stays put until you change it. In addition, if not sure exactly what frequency you want to look at, you can perform a sweep function to look at the SWR or impedance over a wide range of frequencies and then zero in on the frequency that needs the most attention.

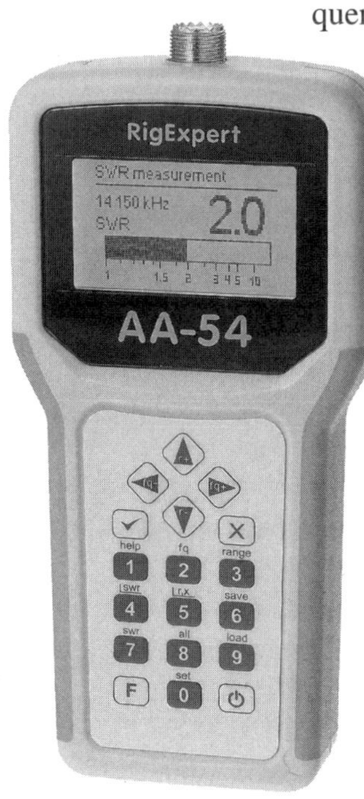

Figure 9.23 — The AA-54 is a very different type of unit than the other three analyzers in this report, but with considerable functional overlap.

On the Bench

Our AA-54 showed remarkable frequency accuracy and setability. At 10 MHz, we found the frequency accuracy to be within 800 Hz of the displayed frequency, quite appropriate for its 1 kHz resolution, expanding to be within 6.5 kHz at 54 MHz. It also stayed on frequency, exhibiting virtually no drift throughout our testing. The total output level was +12.3 dBm, ± 0.1 dBm, over the entire operation range, although there was high harmonic and spurious content in the output. This did not seem to cause any problems with impedance measurements, perhaps due to internal processing, but could make for confusion if being used as a signal generator.

The measured output level of the desired signal component ranged from +12.3 dBm at 100 kHz to +11.5 dBm at 10 MHz. From 15 to 30 MHz it ranges from +1.5 to +2.5 dBm, while from 35 to 54 MHz it is in the –2.0 to –2.5 dBm range. A summary of ARRL Lab testing results is provided in **Table 9.10**.

Operator interaction is provided through a custom key pad and monochrome LCD screen. A single UHF connector on the top goes to the test sample, and a socket for a USB printer type cable is provided for connection to a PC if desired. For most functions, the PC is not necessary, however, while connected, the AA-54 is powered via the USB port.

Table 9.10
Rig Expert AA-54 HF Antenna Analyzer

Manufacturer's Specifications	*Measured in the ARRL Lab*
Frequency range: 0.1-54 MHz.	1.5-54 MHz (usable range).
SWR measurable range: 1:1-10:1	As specified.
Impedance range: 0-1000 Ω.	As specified.
Impedance accuracy: Not specified.	See Table 9.13.
Output power: 20 mW (+13 dBm), 50 Ω load.	17 mW (+12.3 dBm) see text, 50 Ω (1.5-54 MHz).
Power requirements: Two 1.5 V alkaline AA batteries, two 1.2 V NiMH AA batteries, or external power via USB port.	Measurement mode: 244 mA backlight, on, 169 mA backlight off; standby, 60 mA (backlight off), all at 3 V dc.

Size (HWD): 8.5 × 3.8 × 1.5 inches (incl protrusions); weight, 14 oz (with batteries).

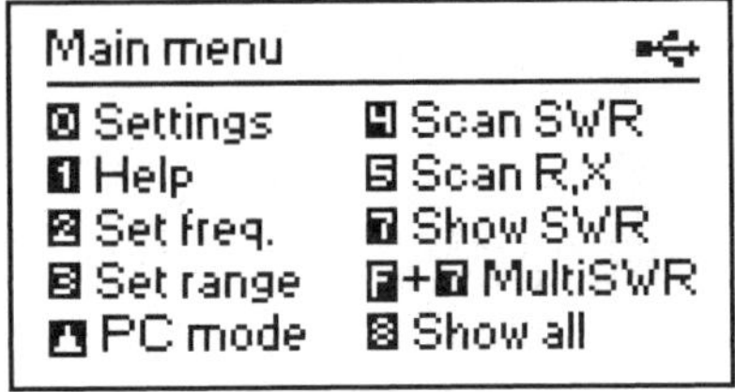

Figure 9.24 — Main menu screen of AA-54.

The AA-54 is menu driven (see **Figure 9.24**) and can provide bar type graphs or numerical SWR or Z data at one or more frequencies. It can also provide swept frequency data. In operation, I found the bar graphs best for making adjustments, since the display updates rapidly, while the swept data is most useful for a summary of results across a band following adjustment or repair.

In the Field

Standing wave ratio (SWR) is the meat and potatoes of such a device. Set for a single frequency, the LCD display shows a nice-to-tune-with calibrated bar graph, a large font SWR readout, to three digits, and the selected frequency, lest you forget (see **Figure 9.25**).

Selecting SHOW ALL on the menu provides the details of the impedance being measured. This provides the measurement frequency, the SWR and also your choice of a series or parallel equivalent model of R and X, including sign and even the calculated equivalent capacitance or inductance value. See **Figure 9.26**. This is much more useful data than just SWR if you wish to design a network to match the load, for example.

Figure 9.25 — Calibrated bar graph display in single frequency mode. The bars respond almost as quickly as an analog meter making them appropriate for antenna tuner or adjusting the controls of an antenna tuner for minimum SWR.

Graphing Modes: A plot of SWR (**Figure 9.27**) or R ± jX (**Figure 9.28**) versus frequency can be easily arranged, again in either series or parallel equivalent model. Unlike many devices the sign of the reactance is shown, as well as its value.

MultiSWR Mode: A nice feature of all these units

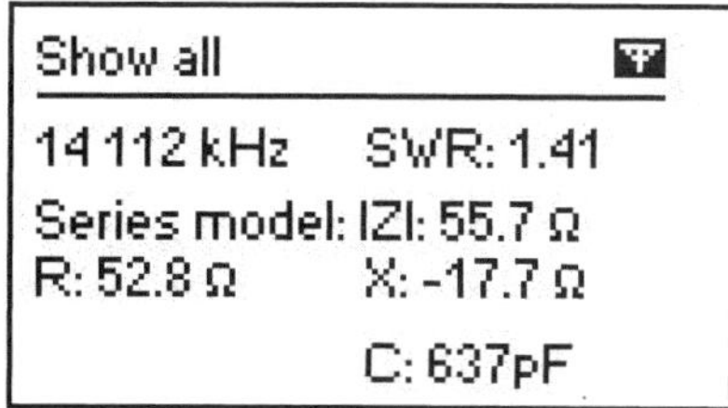

Figure 9.26 — Screen shot of SHOW ALL display in series equivalent mode. A parallel equivalent circuit model may also be selected. Note that the sign of the reactance is provided, along with the equivalent capacitance at the selected frequency.

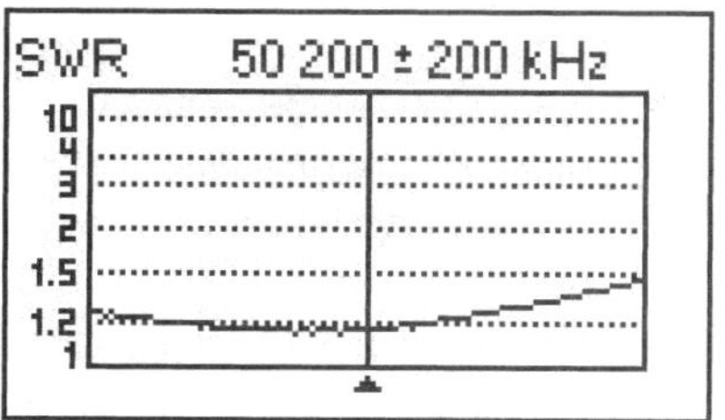

Figure 9.27 — Plot of SWR versus frequency.

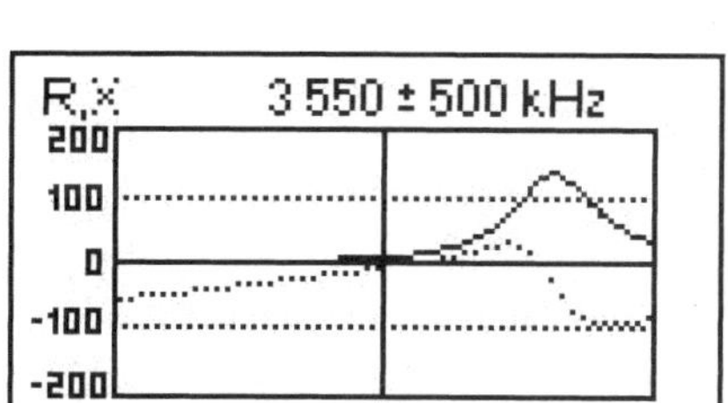

Figure 9.28 — Plot of R ± *j*X versus frequency. Note that the sign of the reactance is shown.

MultiSWR

▶ 14 150 kHz
21 200 kHz
28 300 kHz
50 200 kHz
10 000 kHz

Figure 9.29 — Multi-frequency SWR plot with relative bar graph display. The five frequencies can be anywhere within the meter's range. Very handy for adjusting multiband antennas, especially if they interact.

MultiSWR

▶ 14 150 kHz SWR: 1.5
21 200 kHz SWR: 1.46
28 300 kHz SWR: 1.48
50 200 kHz SWR: 1.18
10 000 kHz SWR: 11.8

Figure 9.30 — Multi-frequency SWR plot with numerical SWR indications on each of five frequencies.

except the AA-30 is that data on multiple distinct frequencies can be observed simultaneously. This can be very useful while making adjustments on multiband antennas. In this case the display shows the frequency and a relative bar graph for each frequency (**Figure 9.29**) or the actual numerical SWR value (**Figure 9.30**). Without this feature, one often has to cycle through the interacting bands multiple times to get them all right. With the AA-54, you can observe the effects on five bands while you make adjustments.

The AA-54 includes a memory capability so that you can store up to 100 display screens. As you store each, you are prompted to tag them with an ID to make sorting them out later easier. They can also be shifted to a PC, great for "as built" or "as adjusted" records for later comparison to see degradation occurring, or confirm it hasn't.

Computer Connection

The AA-54 comes with a CD that includes two auxiliary programs, described below. The software manual indicates that it can be installed in a PC running *Windows 2000, 2003, XP, Vista or 7* as well as *Mac OS* (version 10.6 recommended). I tried it on *Windows XP* and *Windows 7* machines at my location, and each ran successfully.

The disc sets up two programs, *LCD2Clip*, which brings screen shots from the AA-54's display directly into the PC (push F and 6 simultaneously on the AA-54 keypad) at which point you can make screen shots to save the screen with your favorite photo program or *Windows Paint*. That is how Figures 9.24 through 9.30 were obtained.

The other program is a more interesting for many applications. *AntScope* shows results on a full size PC screen, rather than on a copy of the AA-54's small native display. This allows viewing results in more detail, but each does take a few moments to display and transfer data. This program operates with the AA-54 in PC mode, so all definition and operation take place from the PC.

The maj or functions are similar to the AA-54's — all manner of impedance related data can be displayed—SWR, Z and R + *j*X (with sign of X, see **Figure 9.31**). The frequency limits can be set from the PC to display any portion of the range up to 54 MHz wide. By moving the curser with the mouse, all the details can be shown at any selected frequency.

A rather dramatic departure from the typical antenna analyzer is a time domain reflectometer (TDR) function. This sends a virtual pulse down the line and graphically displays the reflected pulse from any discontinuity along the line. The discontinuity could be an antenna at the end of the line, but of even more interest are any and all discontinuities between the source and the antenna, likely indicating a cable fault. While the default display goes out to a distance of 900 meters (probably more useful for a telephone company than the typical amateur operator), it can be reduced to show closer indications as shown in

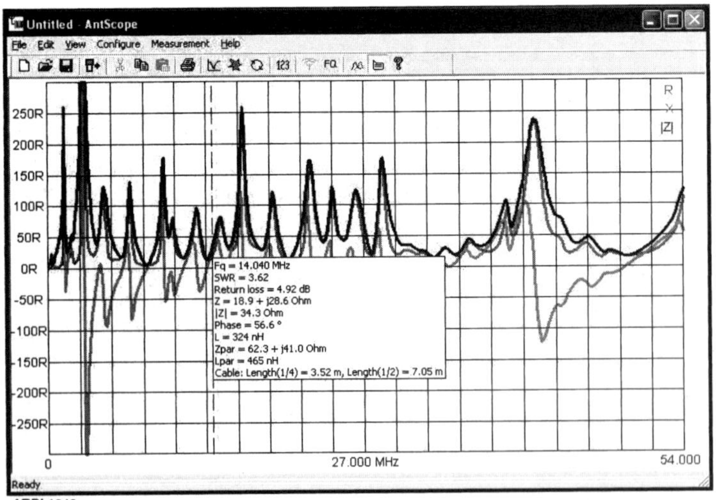

Figure 9.31 — *Antennascope* view of the SWR of an antenna over the entire range. Note that European amateur bands are highlighted. Another view provides R, Z and X. The range can be reduced for more detail over a band, for example. Smith chart views are also provided.

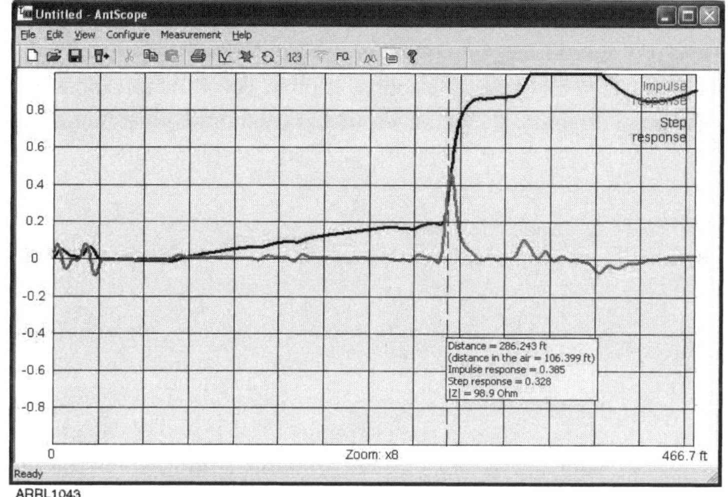

Figure 9.32 — *Antennascope* in time domain relectometer (TDR) mode. The TDR provides a radar-like view of cable discontinuities along the line. Here you see the 280 foot run of 0.82 relative velocity coax to the W1ZR 80 meter ground plane antenna. The early blips are the pulse partially reflecting from the impedance bumps going through my bypassed linear amplifier and antenna tuner. At almost three divisions (about 80 feet) out, the pulse encounters my dc grounded lightning arrestor at the entrance panel and then 200 feet of coax to the antenna. The details are shown by mousing the cursor to the discontinuity. Had there been a break, short or other cable problem, this would show you exactly where it was.

Figure 9.32. In this view it has been changed to use US metrics. This feature is something usually found in much more expensive instruments and has the potential to be a great diagnostic tool. The manufacturer notes that it is really intended for the AA-230 and higher frequency units that provide additional resolution, but can be used with the lower resolution AA-54 to discover major discontinuities as shown.

Documentation

The AA-54 is provided with a 22 page *User's Manual*, also available on their website if you want to look it over before you buy. The manual does a good job of describing the basic functions of the device. In addition, the last eight pages are devoted to using the AA-54 in various applications. This section starts with antennas, but moves through measuring characteristics of cables, lumped inductors and capacitors and transformers. The use as an RF signal generator is also covered, with some cautions as to waveform.

The AA-54 also comes with an 11 page *Software Manual*. This describes how to load and run the programs discussed previously. While *LCD2Clip* is very simple to use, *AntScope* offers many features and adjustments. The choices on each tab are shown in pages 8 and 9 of the manual, but you will likely need to try each to see what they are about. I had no trouble installing, running or using the supplied software.

Manufacturer: Rig Expert Ukraine Ltd, Oranzhereyna 3, 04112 Kiev, Ukraine; e-mail: **info@rigexpert.com**; **www.rigexpert.com**.

Bottom Line

The AA-54 is a very competent, accurate and easy to use analyzer providing single or multi-frequency pointed or plotted SWR and impedance data on a useful LCD display. In the shack or lab, it can also provide more advanced features while connected to a PC using the supplied software.

YouKits FG-01 Antenna Analyzer

The YouKits FG-01 analyzer (sold in the US by TEN-TEC, see **Figure 9.33**) is the most compact of the bunch — not a lot bigger than a pack of cigarettes, if I remember them correctly. It is also the least expensive of this group, although it does require an optional battery pack to be self contained. It measures SWR and magnitude of impedance from 1 to 60 MHz, showing the numerical result of each at the chosen center frequency along with a display of the swept frequency results, all on a small but readable color display screen.

The FG-01 is a very easy to operate unit. There is a single knob which, by default sets the center frequency of the analysis, also the frequency that the numerical data applies to. The tuning is over a single continuous band with the tuning step size set from 1 MHz to 1 kHz in four steps. The steps are selected by pushing in the knob for 1 second. The digit that will be changed flashes and the digit can be changed by turning the knob while holding it in. Once the step size is selected, the tuning will be at that step size until you change it. After it is set up the way you want it, pushing down the knob for 1 second will save your settings for the next time you power it up.

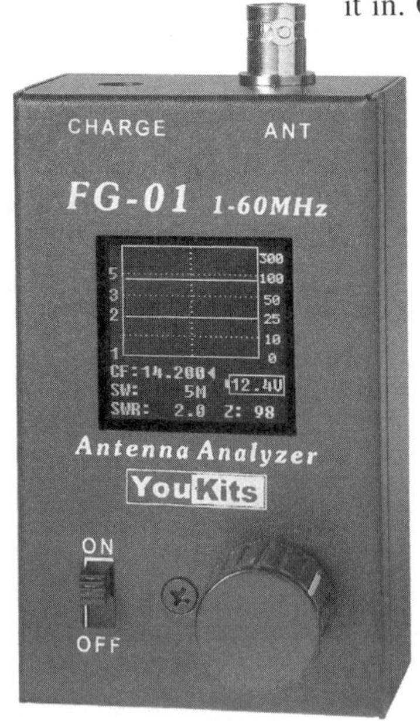

Figure 9.33 — The YouKits FG-01 analyzer is not a lot bigger than a pack of cigarettes,

I found the color display easy to see with one exception. It was sometimes difficult for me to decide which of the two plots was which, since the colors don't seem that far apart. Fortunately, the manufacturer seems to have anticipated this. If you hold the TUNING knob in while you switch the unit on, it will just plot the SWR, the most useful information for most applications. The impedance is still shown in the numerical data portion of the display.

The other aspects of the display are easy to use. It simultaneously displays the center frequency, sweep width, SWR, impedance magnitude (no information on the complex impedance, as with the Comet) and battery voltage. The battery display turns red if the voltage drops below 9.5 V and the SWR changes to red for an SWR of greater than 3:1.

On the Bench

The summary of lab measurements shown in **Table 9.11** reflect a very competent instrument. In addition it was noted that the frequency stayed unchanged once set. Of course, many of the others were quite stable over the frequency range of the FG-01 still, it does a very good job at what it does. The spectral purity was the best we saw (see **Figure 9.34**).

Documentation

The unit is supplied with a five page instruction pamphlet that covers the basic operational details, along with many caution notices. It covers how to work it, apparently with the idea that if you buy one, you already know why

Table 9.11
YouKits FG-01 SWR/Impedance Analyzer

Manufacturer's Specifications	*Measured in the ARRL Lab*
Frequency range: 1-60 MHz.	As specified.
SWR measurable range: Not specified.	1.0-9.0:1
Impedance range: Not specified.	5-350 Ω.
Impedance accuracy: Not specified.	See Table 9.13.
Output power: 32 mW max (+15 dBm),	36 mW (+15.5 dBm) into 50 Ω at 14 MHz. 23 mW (+13.6 dBm) into 50 Ω at 50 MHz.
Power requirements: 400 mA at 10-12.8 V dc.	398 mA at 12.8 V dc (external power); 379 mA at 12.4 V dc (internal batteries).

Size (HWD): 4.4 × 2.3 × 2.2 inches, (incl protrusions); weight: 13.5 oz with internal battery.

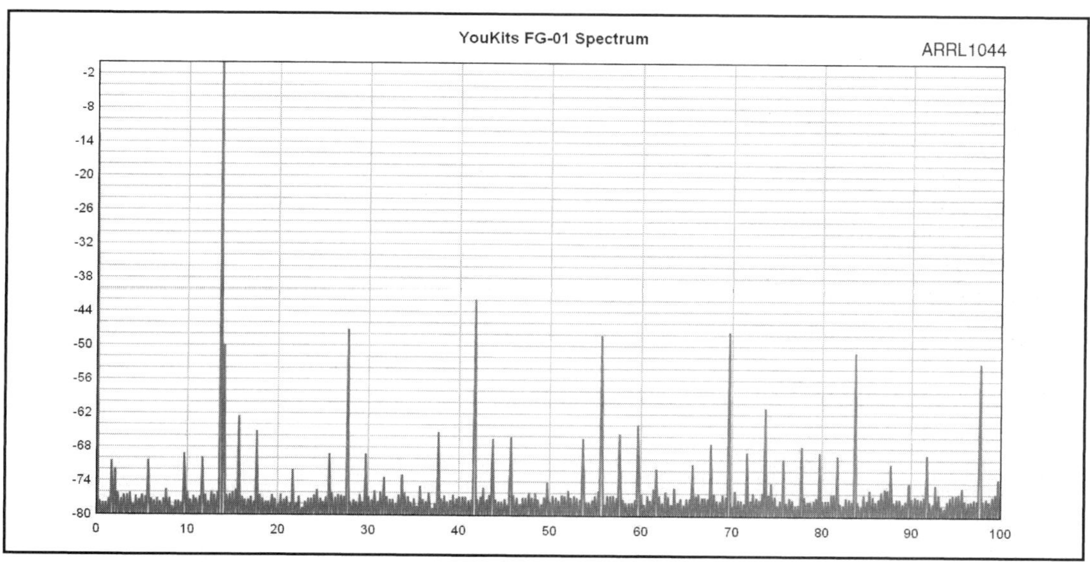

Figure 9.34 — Output spectrum of the YouKits FG-01 while tuned to 14 MHz. The cleanliness of the spectrum is notable, sufficient that the unit could be used as a very low power (QRPp) transmitter, if there were a way to key it.

you wanted it and what you can do with it. The instructions are available on the TEN-TEC website for a preview.

One caution relates to power sources. You are cautioned that the AA size battery holder provided is only for 3.6 V lithium cells, rather than the usual 1.5 V alkaline cells. It also notes that if an external supply is used, it must provide 10 to 12.8 V dc, letting out most station dc systems.

Manufacturer: YouKits, sold in the US by TEN-TEC, 1185 Dolly Parton Parkway, Sevierville, TN 37862; tel 800-833-7373; e-mail **sales@tentec.com**; **www.tentec.com**.

Bottom Line

This unit shares the measurement capabilities of the Comet, but has a digital rather than analog display, and adds the handy sweep function. The frequency stability and accuracy are notable. It is easy to operate, easy to carry, compact and does what it does quite nicely.

A Comparison of the Units in the Group

Table 9.12 illustrates the features of the four units side by side for comparison. In addition, a detailed tabulation of the impedance measurements of each unit is shown in **Table 9.13.**

Notes

[1]J. Hallas, W1ZR, "A Look at Four Antenna Analyzers," Product Review, *QST*, Mar 2013, pp 46-52.

Table 9.12
Antenna Analyzer Feature Comparison

Analyzer	*Price*	*Range (MHz)*	*SWR*	*Z*	*X*	*Sign of X*	*AA Batt*	*Ext. Pwr*	*Socket*	*PC I/O*	*Storage Locations*
Comet CAA-500	$420	1.53-259 273-508	1-6	Yes	No	N/A	6	Yes	UHF/N*	No	No
MFJ-266	$320	1.5-71 85-185 300-490	1-9.9	Yes	Yes	No	8	Yes	N**	No	No
RigExpert-AA-54	$320	0.1-54	1-10	Yes	Yes	Yes	2	Yes	UHF	Yes	100
YouKits FG-01	$249	1-60	1-9	Yes	No	No	3***	Yes	BNC	No	No

*Type N for top frequency range only.
**Type N to UHF adapter provided.
***Requires 3.6 V lithium batteries, type 14500.

Table 9.13
Impedance and SWR Measurements

Load	*Frequency*	*Comet CAA-500*	*MFJ -266*	*RigExpert AA-54*	*YouKits FG-01*	*Agilent 4291B (reference)**
50 Ω (1:1 SWR)	3.5 MHz	50 Ω (1.0:1)	50+*j*0 Ω (1.0:1)	49.8–*j*0.2 Ω (1.0:1)	48 Ω (1.0:1)	50+*j*0 Ω[2]
	14 MHz	50 Ω (1.0:1)	50+*j*0 Ω (1.0:1)	49.8+*j*1.1 Ω (1.0:1)	48 Ω (1.0:1)	50+*j*0 Ω
	28 MHz	50 Ω (1.0:1)	50.0+*j*0 Ω (1.0:1)	49.8+*j*2.2 Ω (1.0:1)	48 Ω (1.0:1)	50+*j*0 Ω
	50 MHz	50 Ω (1.0:1)	49+*j*0 Ω (1.0:1)	49.7–*j*4.0 Ω (1.1:1)	48 Ω (1.0:1)	50+*j*0 Ω
	144 MHz	50 Ω (1.0:1)	49–*j*0 Ω (1.0:1)	—	—	50+*j*0 Ω
	223 MHz	48 Ω (1.1:1)	—	—	—	50+*j*0 Ω
	440 MHz	50 Ω (1.1:1)	— (1.1:1)	—	—	50+*j*0 Ω
5 Ω (10:1 SWR)	3.5 MHz	—	4+*j*3 Ω (>9.9:1)	5.0+*j*0.4 Ω (9.9:1)	4 Ω	5.1+*j*0.0 Ω (8.5:1)
	14 MHz	—	2+*j*4 Ω (>9.9:1)	5.1+*j*1.7 Ω (9.9:1)	4 Ω	5.1+*j*0.2 Ω (8.5:1)
	28 MHz	—	3+*j*4 Ω (>9.9:1)	5.1+*j*3.4 Ω (9.8:1)	6 Ω	5.1+*j*0.4 Ω (8.7:1)
	50 MHz	—	6+*j*0 Ω (7.3:1)	5.1+*j*5.8 Ω (9.7:1)	11 Ω	5.1+*j*0.7 Ω (8.2:1)
	144 MHz	—	3+*j*9 Ω (>9.9:1)	—	—	5.2+*j*1.9 Ω
	440 MHz	—	(>9.9:1)	—	—	—
25 Ω (2:1 SWR)	3.5 MHz	25 Ω (1.8:1)	25–*j*0 Ω (2.0:1)	25.1+*j*0.1 Ω (2.0:1)	23 Ω	25.1+*j*0 Ω (1.9:1)
	14 MHz	25 Ω (1.8:1)	23+*j*12 Ω (2.2:1)	25.1+*j*0.7 Ω (2.0:1)	24 Ω	25.1+*j*0.2 Ω (1.9:1)
	28 MHz	26 Ω (1.8:1)	24+*j*10 Ω (2.1:1)	25.2+*j*1.3 Ω (2.0:1)	24 Ω	25.1+*j*0.4 Ω (2.0:1)
	50 MHz	26 Ω (1.8:1)	26+*j*0 Ω (1.8:1)	25.2+*j*2.4 Ω (2.0:1)	25 Ω	25.1+*j*0.7 Ω (1.9:1)
	144 MHz	24 Ω (1.7:1)	26+*j*10 Ω (2.0:1)	—	—	25.2+*j*2.0 Ω
	223 MHz	32 Ω (1.5:1)	—	—	—	—
	440 MHz	40 Ω (1.9:1)	— (1.5:1)	—	—	—
100 Ω (2:1 SWR)	3.5 MHz	110 Ω (1.9:1)	95–*j*19 Ω (2.0:1)	99.6+*j*1.7 Ω (2.0:1)	98 Ω (2.0:1)	100–*j*0.2 Ω
	14 MHz	110 Ω (1.9:1)	90–*j*33 Ω (2.1:1)	99.0+*j*8.8 Ω (2.0:1)	98 Ω (2.0:1)	100–*j*0.9 Ω

[continued]

Load	*Frequency*	*Comet CAA-500*	*MFJ -266*	*RigExpert AA-54*	*YouKits FG-01*	*Agilent 4291B (reference)**
	28 MHz	105 Ω (1.9:1)	89–*j*33 Ω (2.1:1)	97.0+*j*16.8 Ω (2.0:1)	98 Ω (2.0:1)	100–*j*1.8 Ω
	50 MHz	100 Ω (1.9:1)	90–*j*27 Ω (2.0:1)	92.0+*j*28.6 Ω (2.1:1)	95 Ω (1.9:1)	99.9–*j*3.1 Ω
	144 MHz	92 Ω (2.1:1)	74–*j*41 Ω (2.1:1)	—	—	99–*j*8.9 Ω
	223 MHz	80 Ω (2.4:1)	—	—	—	—
	440 MHz	90 Ω (2.0:1)	— (2.1:1)	—	—	—
200 Ω (4:1 SWR)	3.5 MHz	225 Ω (3.8:1)	160–*j*94 Ω (4.4:1)	197.8–*j*1.7 Ω (4.0:1)	205 Ω (4.0:1)	201–*j*1.2 Ω
	14 MHz	225 Ω (3.8:1)	149–*j*107 Ω (4.6:1)	195.0–*j*26.1 Ω (4.0:1)	205 Ω (4.0:1)	201–*j*4.8 Ω
	28 MHz	220 Ω (3.8:1)	144–*j*104 Ω (4.5:1)	187.5–*j*50.3 Ω (4.0:1)	205 Ω (4.0:1)	200–*j*9.4 Ω
	50 MHz	210 Ω (3.7:1)	132–*j*100 Ω (4.3:1)	164.2–*j*83.4 Ω (4.2:1)	195 Ω (4.0:1)	199–*j*16 Ω
	144 MHz	175 Ω (4.1:1)	72–*j*93 Ω (4.3:1)	—	—	189–*j*45 Ω
	223 MHz	125 Ω (4.8:1)	—	—	—	—
	440 MHz	170 Ω (4.1:1)	— (3.8:1)	—	—	—
1000 Ω (20:1 SWR)	3.5 MHz	—	—	883–*j*184 Ω (18.7:1)		998–*j*33 Ω
	14 MHz	—	—	505–*j*471 Ω (18.6:1)		981–*j*127 Ω
	28 MHz	—	—	202–*j*471 Ω (∞)		935–*j*230 Ω
	50 MHz	—	—	53–*j*270 Ω (∞)		825–*j*373 Ω
50 – *j*50 Ω (2.62:1 SWR)	3.5 MHz	70 Ω (2.5:1)	34–*j*39 Ω (2.6:1)	49.0–*j*46.1 Ω (2.5:1)	83 Ω (2.5:1)	50–*j*47 Ω
	14 MHz	75 Ω (2.8:1)	33–*j*51 Ω (3.5:1)	45.5–*j*51.5 Ω (2.8:1)	89 Ω (2.7:1)	48–*j*52 Ω
	28 MHz	70 Ω (2.5:1)	36–*j*45 Ω (2.8:1)	45.8–*j*46.9 Ω (2.6:1)	78 Ω (2.4:1)	51–*j*48 Ω
50 + *j*50 Ω (2.62:1 SWR)	3.5 MHz	80 Ω (2.6:1)	65+*j*54 Ω (2.6:1)	52.0+*j*50 Ω (2.6:1)	92 Ω (2.5:1)	52+*j*50 Ω
	14 MHz	75 Ω (2.5:1)	51+*j*51 Ω (3.0:1)	55.8+*j*48.1 Ω (2.4:1)	92 Ω (2.5:1)	53+*j*48 Ω
	28 MHz	90 Ω (2.5:1)	54+*j*59 Ω (2.9:1)	72.2+*j*49.0 Ω (2.4:1)	100 Ω (2.5:1)	65–*j*51 Ω

*The SWR loads constructed in the ARRL Lab were measured on an Agilent 4291B Impedance Analyzer by ARRL Technical Advisor John Grebenkemper, KI6WX. An HP 11593A precision termination was used for the 50 Ω tests. This termination has a wide frequency range.

INDEX

Note: The letters "ff" after a page number indicate coverage of the indexed topic on succeeding pages.

A

We're interested in hearing your comments on this book and what you'd like to see in future editions. Please email comments to us at **pubsfdbk@arrl.org**, including your name, call sign, email address and the title, edition and printing of this book. Or please mail comments to ARRL, 225 Main St, Newington, CT 06111-1494.